SONNY MUIR.

A GUIDE TO THE
REPTILES
AND
FROGS
OF THE
PERTH
REGION

ABOUT THE AUTHORS

The authors have all had a passionate interest in reptiles and frogs since childhood. They met through the Western Australian Naturalists' Club, and their detailed knowledge of the subject has been accumulated over many years of amateur study. They have published numerous articles on herpetology, and were founder members of the Western Australian Society of Amateur Herpetologists. They also operate a voluntary snake-removal service.

Brian Bush worked as a fencer in Esperance and the Eastern Goldfields for 10 years, pursuing his fascination with reptiles. Since 1986, he has been working in Perth in special education on venomous animals, especially snakes.

Brad Maryan has been involved with numerous reptile surveys and studies in Western Australia. He has been working at the Perth Zoo since 1987.

Robert Browne-Cooper was 13 years old when he caught his first snake. Membership of the WA Naturalists' Club fuelled his interest and gave him opportunities to study and contribute to knowledge of reptiles and amphibians. He now works as a Laboratory Technician.

David Robinson has been fascinated by reptiles for as long as he can remember. He is a field worker with the Agricultural Protection Board.

A GUIDE TO THE REPTILES AND FROGS OF THE PERTH REGION

■ BRIAN BUSH ■ BRAD MARYAN ■
■ ROBERT BROWNE-COOPER ■ DAVID ROBINSON ■

UNIVERSITY OF WESTERN AUSTRALIA PRESS

First published in 1995 by
University of Western Australia Press
Nedlands, Western Australia, 6907.

This book is copyright. Apart from any fair dealing for the purpose of private study, research, criticism or review, as permitted under the Copyright Act 1968, no part may be reproduced by any process without written permission. Enquiries should be made to the publisher.

© Copyright Brian Bush, Brad Maryan, Robert Browne-Cooper, David Robinson 1995

National Library of Australia
Cataloguing-in-Publication entry:

A guide to the reptiles & frogs of the Perth region.

   Bibliography.
   Includes index.
   ISBN 1 875560 42 4.

   1. Reptiles—Western Australia—Perth Region—
   Identification. 2. Frogs—Western Australia—Perth
   Region—Identification. I. Bush, Brian, 1947- .
   II. Title: Guide to the reptiles and frogs of the Perth region.

597.9099411

Consultant Editor: Kate Hooper, Perth
Designed by Robyn Mundy, Mundy Design, Perth
Drawings on pages 5, 24, 60, 109, 145 & 226 by Neil Elliott
Typeset in 10pt Century Old Style by Lasertype, Perth
Printed by Scott Four Colour Print, Perth

Dedicated to

Dr Glen Milton Storr
(1921-1990)

His passing marked the end of an exciting era in Australian herpetology. In particular, it was an era of great progress in the taxonomic classification of Western Australia's reptiles. During his 28 years as Curator of Herpetology and Ornithology at the Western Australian Museum, he formally described three genera, 180 species and 50 subspecies of lizard and snake.

Glen was a very special person to know and an inspiration to any naturalist fortunate enough to cross his path. He always encouraged the efforts of amateurs and we came to know him as 'the awesome herpetologist', not in reference to his size but for his enthusiasm and love for natural history.

Contents

Preface	ix
Acknowledgments	xi
Western Australian Society of Amateur Herpetologists	xiii

Introduction — 1
Using this book	3
Habitats	7
Conservation	16

Frogs — 21
ground frogs (Myobatrachidae)	27
tree frogs and water-holding frogs (Hylidae)	43

Sea Turtles — 46

Freshwater Turtles — 49

Lizards and Snakes — 54
What do lizards and snakes have in common?	55
What are the differences between lizards and snakes?	58

Lizards — 60
geckos (Gekkonidae)	62
legless lizards (Pygopodidae)	79
dragon lizards (Agamidae)	92

goannas or monitor lizards (Varanidae)	99
skinks (Scincidae)	106

Snakes 143

Land snakes 145
    blind or worm snakes (Typhlopidae) 149
    pythons (Boidae) 156
    front-fanged venomous land snakes (Elapidae) 162

Sea snakes (Hydrophiidae) 192

Is it true that...? 195

Snakebite 198

    snakes and snakebite 198
    seasons and snakebite 200
    first aid for snakebite 202, 225

Snakes and People 203

Dogs, Cats and Snakes 204

    precautions 204
    initials symptoms of snakebite in dogs and cats 205

Bibliography	207
References	209
Glossary	211
Index	220

Preface

Reptiles and frogs are fascinating animals, yet most people have little to do with them. They are often looked upon with abhorrence and fear, especially the larger crocodiles and venomous snakes.

The study of reptiles and amphibians is called herpetology. When we decided to put this guide together we were aware of the large number of excellent books available on this subject. But most of these are comprehensive works, and because of the diversity of the Australian herpetofauna, they are large and often quite expensive. Books of a regional type such as this one have the advantage of giving you specific information on the reptiles and frogs that inhabit your area, at less cost.

When people think of reptiles they often think of the dinosaur era, but the supremacy of the reptiles did not end there. It may surprise you to know that the reptiles are the most successful and diverse group of terrestrial vertebrates existing today. There are more than 6000 species of lizard and snake in the world; twice the number of species of mammal.

The Perth region has its share of reptiles and frogs, but they are rarely seen. Even when they are, most people would not be able to distinguish a Gwardar from a Dugite. This book is for the student of herpetology and the backyard gardener, the biologist and the bushwalker and all those with an interest in natural history. We hope it will help you to name that slithering, scampering or croaking critter in the backyard, workplace, park, golf course or bush. Maybe this book will also give you an insight into their natural history. Good luck!

Acknowledgments

Special thanks go to Otto Mueller for his critical review of the manuscript during preparation. Any errors or omissions still to be found are of our doing. Thanks also to Greg Harold, Harry Ehmann, Gerald Kuchling and Paul Orange for providing photographs, to Graeme Thompson for allowing us to photograph *Varanus rosenbergi* in his care, to Martin Clery for preparing the regional map, to Jodie Gibson for her assistance with distribution maps and to Michael Browne-Cooper and Geoff Bartlet for their assistance with word processing during the early stages of manuscript preparation. For their support and encouragement we thank John Dell, Ric How, Laurie Smith, Ron Johnstone, Ken Aplin, Betty Wellington and Mark Cowan at the WA Museum. We must also thank the 'phantom' referee. His or her herpetological knowledge and commitment to the task are commendable.

Much of our field work has been funded by a Harry Butler Grant administered by the WA Museum, and carried out under licences issued by the Department of Conservation and Land Management.

Western Australian Society of Amateur Herpetologists

Natural history is a subject that enthrals many people. Although some have a general interest, many are drawn to a more specific field of study. This could be the flora generally, or specifically trees, wildflowers or fungi. Others are drawn to the mammals, birds or fishes. One of us has a friend who is enthusiastically involved in the study of molluscs: 'slimy little beasts that sometimes live in shells'. We could not think of anything less interesting, but then our affinity is with reptiles. The enthusiasm he displays when talking about molluscs is no less than that displayed by us reptile-lovers. And with no more than a passionate interest, he has made an important contribution to knowledge of molluscs. For some people, their interest in natural history is so strong that they can no more live without it than live without oxygen. It may even be suggested that this obsession is determined genetically!

In Australia, amateur natural historians have made a large contribution to our knowledge of the native flora and fauna. There are formal societies and clubs for those involved with reptiles. These allow reptile-lovers to meet others of similar ilk, share information and offer encouragement and direction to the next generation of interested youngsters.

The Western Australian Society of Amateur Herpetologists (WASAH) was initiated informally in November 1990. Our initial aims were to lobby the relevant government ministers and negotiate with the Department of Conservation and Land Management to get amateur herpetology recognised as a hobby in Western Australia. We have succeeded in this.

WASAH is now formally recognised as a society of people with a common interest in herpetology, who share information and experiences. For further information, please contact Brian Bush on (09) 295 3007.

Introduction

PEOPLE generally either like or dislike reptiles, and the majority of people probably dislike them. Frogs, on the other hand, have a more cute and friendly appearance and, although somewhat slimy, are not looked upon as being harmful. Frogs, turtles, lizards and snakes are all an integral part of the Australian bush. As more people become aware of the need to conserve our wildlife heritage there is a greater acceptance of, and interest in, reptiles and frogs. Australia has an extensive and diverse frog and reptile fauna (herpetofauna). It is inevitable that people will encounter these creatures during their recreational and working activities, but they are likely to have problems identifying them.

In the Perth region you may come across any of 16 different frogs, two freshwater turtles, 51 lizards or 24 snakes that occur naturally here. Because the marine reptiles are rarely seen, we decided to concentrate less on these and more on the terrestrial varieties. If you cannot identify a marine reptile using this guide, or you require more information than is included here, look at the bibliography towards the end of this book or contact the Western Australian Museum.

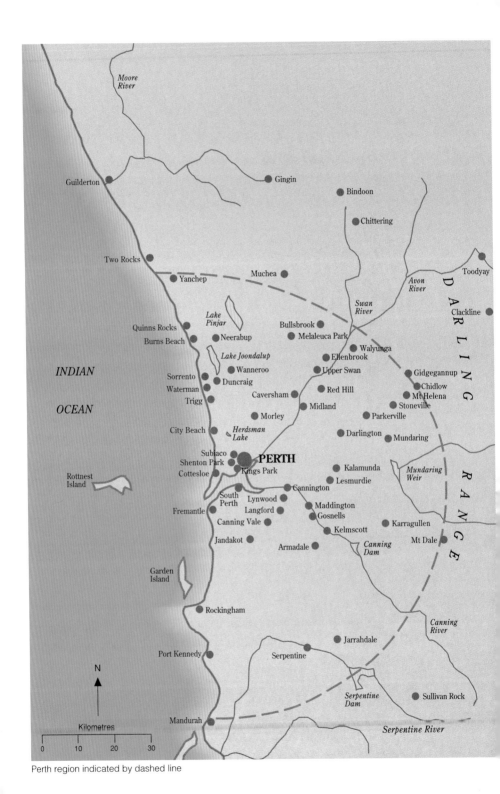

Perth region indicated by dashed line

Using this book

The main purpose of this book is to enable you to identify any reptile or frog you might come across. This should be achievable using a combination of the 'keys', descriptions and, most importantly, the photographs. Obviously, some species are more easily identified than others. Most people would have little trouble identifying a Bobtail Skink, but many other reptiles and frogs are difficult to identify past the generic level. We hope this guide will also be used as a source of general information, by readers ranging from the professional biologist and amateur naturalist to the school student.

In the case of the larger and more noticeable reptiles, the photographs included here are probably sufficient to confirm an identification. However, many frogs and reptiles are small secretive creatures and are not often encountered. These are the ones most often brought to us for identification after being found in the house, killed by the cat, uncovered during gardening or discovered in the woodheap. To identify these you may have to consult the 'keys' as well as the text. Keys are included to allow you to progress from any level of classification to identifying a particular species.

What is a key?

A key is a series of numbered couplets of contrasting character states. Keys are used in the following way:

> Start at the first pair of descriptions (1). The animal being examined will conform with one description or the other. On the right-hand side, adjacent to the description that

suits the animal, you will find either its name or a number. If you find a number, move to the couplet bearing that number, skipping those in between if necessary. Decide which of these two descriptions best suits the animal. If this leads you to another number, repeat the process until eventually you have a name.

If all goes well and the animal being identified is not too atypical, you will have the correct name for it. Proceed to the photograph and double-check that your animal looks similar, then move to the species description.

What are scientific names?

The two-word italicised names found following the common names are scientific names. The first word is the genus (plural, *genera*), which begins with a capital letter, and the second word is the species (plural, *species*). If there is a recognised subspecies or race, then there may be a third italicised word. The name of the person or people who described a genus or species for the first time will be found in upright letters after the scientific name, followed by the year the description was formally published. If the discoverer's name is enclosed in brackets, then the species has since been placed in a different genus to that originally assigned.

As we learn more about Australia's reptiles and frogs, there are changes to the scientific classifications. These changes can cause some confusion. In this book we have generally followed Cogger (1992) in naming the species, with a few exceptions. Where these occur we have included a reference to the first use of the scientific name in other literature.

Photographs

Where possible, we have used photographs of typical local animals, but in some cases we have had to include a photograph of an animal from outside this region. In many species colour can be extremely variable. For these we have included photographs illustrating some of the variations. There is also colour variation between sexes in some species. This is known as sexual colour dimorphism. Where possible, we have included photographs of both sexes. Although this guide will, in most cases, allow correct identification, there may be individual frogs and reptiles that cannot be identified. In these cases you may want to contact the Western Australian Museum.

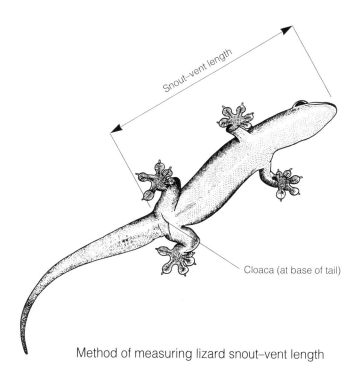

Method of measuring lizard snout–vent length

Unfamiliar terms

While reading this book you will come across numerous words not commonly used in everyday speech, but do not be intimidated by these. There is a large and comprehensive glossary, which will allow you quickly to understand unfamiliar words. The symbol ♂ means male, and ♀ means female. You may also wonder why the measurements included in the descriptions are snout-vent lengths (SVL). This is because many reptiles, particularly lizards, can have damaged or regrown tails, making this appendage an unreliable character for use in identification.

The Perth area

The map on page 2 shows the region covered by this book. It extends north to Yanchep, south to Mandurah, east to the Darling Range and west to the offshore islands in the Indian Ocean. Many of the frogs and reptiles found outside this area can still be identified with this guide, as many locally found species are widespread throughout much of the south-west of Western Australia.

The mainland in this region can be divided into two major land units that contain different assemblages of herpetofauna because they have different soils, vegetation and geography. They are the Swan Coastal Plain and the Darling Range. Through reading this book it will become obvious that species are usually associated with only one of these two areas (although some occur in both). However, many species are dependant on a certain type of soil or vegetation for their survival. Therefore within these broad areas, a species may be distributed in a series of discontinuous populations or patches, which relate to the occurrence of suitable areas in which it can live.

Habitats

A place within an environment where an organism lives is called its habitat. The specific place an animal occupies within one of these areas is its microhabitat. A combination of where it lives, what it does, and its relationship with the plants and other animals in the area, allow it to fill an exclusive ecological niche.

The Perth region has undergone major urban and industrial development. This in turn has caused large-scale fragmentation of habitats. The remnant areas of bushland include national parks, nature reserves and private property. It is within these areas of bushland that you may see most of the habitat types described here. We have avoided technical descriptions of the various vegetation communities and substrates in the region. Instead, we describe the main habitat types on the basis of our experience of the species found within them. Our habitat classifications are meant only as a guide, to enable you to anticipate which reptiles and frogs are likely to occur in a given area or habitat.

Most of the species occurring in the Perth region are widely distributed throughout much of the south-west of Western Australia. A few occur over much of Australia. The Fence Lizard (*Cryptoblepharus plagiocephalus*) and the Burton's Legless Lizard (*Lialis burtonis*) live in a broad range of habitats and, therefore, are found throughout most of this region. These are habitat-general species. In contrast, some species have very particular requirements. They are habitat-specific species. For example, the Reticulated Velvet Gecko (*Oedura reticulata*) is restricted to smooth-barked eucalypt woodlands, and the Ornate Dragon (*Ctenophorus ornatus*) is restricted to granite. Both are common and widespread outside the Perth area, but only in their respective habitats.

Humans getting on with the business of living change things on a comparatively large scale. These changes do not always disadvantage other life forms. In fact, we create new habitats, which are often utilised by the habitat-general species.

Human-made habitats

Anyone involved with reptiles will tell you that rubbish tips are some of the best places to find them in large numbers. The many hiding spots and dense food base allow quite large and diverse populations to survive at these sites. Farms, with their grassland pastures and permanent dams, are ideal for some of the larger lizards and snakes. Spoil-heaps of loose soil and vegetation are caused by earth-moving machines during firebreak or road construction. These are quickly occupied by the smaller burrowing species. Even our backyards will support the more adaptable frogs and reptiles. Blind snakes can be unearthed while gardening. Geckos and skink lizards may be seen on fences and walls. Frogs love rockeries, especially if there is a swimming pool or fish pond nearby. But the majority of species are not so adaptable. The smallest change to the environment in which they live is often all that is needed to cause their demise.

Offshore islands

Rottnest and Garden Islands have been included within the scope of this book. The limestone rocks or islets visible in Shoalwater Bay and Warnbro Sound are excluded, as they support very little vegetation and therefore few, if any, reptiles and frogs. The islands were isolated from the mainland as a result of a post-Pleistocene rise in sea level. Rottnest, the largest local island, was separated as recently as 6000–7000 years ago.

The island flora is similar to that on the adjacent mainland and, because of their size, is most diverse on Rottnest Island and Garden Island. They are large enough to support trees such as the Rottnest Pine (*Callitris preissii*) and Rottnest Teatree (*Melaleuca lanceolata*), although they both lack some of the types of tree that are widespread on the mainland, including banksias, blackboys and eucalypts. Non-native species, along with eucalypts, have been recently introduced to the islands.

The degree of habitat disturbance varies markedly between the two islands. People have limited access to Garden Island because it is an Australian naval base. Hence, there is minimal habitat disturbance. This is reflected in the abundance of Carpet Pythons (*Morelia spilota*) on the island compared with the adjacent mainland.

Coastal limestone and heath, Rottnest Island

A very dry habitat supporting few frog species but many reptiles. A sample of species found in this habitat are the South-western Spiny-tailed Gecko (*Strophurus spinigerus*), West Coast Ctenotus (*Ctenotus fallens*), Two-toed Earless Skink (*Hemiergis quadrilineata*), King's Skink (*Egernia kingii*), Line-spotted Lerista (*Lerista lineopunctulata*), Western Pale-flecked Morethia (*Morethia lineoocellata*) and all three banded burrowing snakes (*Simoselaps* spp.).

Tuart woodland, Waterman

This woodland habitat occurs along a narrow near-coastal strip on the Swan Coastal Plain, on both undulating and level ground. Near Perth it is generally associated with other eucalypts and grows over a mixed shrub and heath understorey supporting banksia and acacia trees.

A sample of species found in this woodland are the Moaning Frog (*Heleioporus eyrei*), Western Banjo Frog (*Limnodynastes dorsalis*), Western Marbled Gecko (*Phyllodactylus marmoratus*), Gray's Legless Lizard (*Delma grayii*), Western Slender Bluetongue (*Cyclodomorphus branchialis*), Western Worm Lerista (*Lerista praepedita*) and the Bardick (*Echiopsis curta*), which resembles an adder.

Banksia woodland, Ellenbrook

An extensive habitat on the Swan Coastal Plain occuring in varying densities. It is usually intermixed with a variety of other small trees such as blackboys (*Xanthorrhoea* spp.), sheoaks (*Casuarina* spp.) and woollybush (*Adenanthos* spp.).

A sample of species found in this woodland are the Turtle Frog (*Myobatrachus gouldii*), South-western Sandplain Worm Lizard (*Aprasia repens*), Keeled Legless Lizard (*Pletholax gracilis*), Western Heath Dragon (*Tympanocryptis adelaidensis*), Jewelled Ctenotus (*Ctenotus gemmula*), Western Limestone Ctenotus (*C. lesueurii*) and the Black-striped Snake (*Neelaps calonotus*).

Jarrah woodland, Chidlow

The dominant woodland habitat on the Darling Range, which includes Marri (*Eucalyptus calophylla*) over a variety of low shrubs and occasionally heath. Bull Banksia (*Banksia grandis*), Zamia Palm (*Macrozamia reidlei*) and Prickly Moses (*Acacia pulchella*) are common understorey growth in here.

A sample of species found in this habitat are the Rusty Gecko (*Diplodactylus polyophthalmus*), Darling Range Heath Ctenotus (*Ctenotus delli*), Southern Five-toed Earless Skink (*Hemiergis initialis*), South-western Four-toed Lerista (*Lerista distinguenda*), Southern Pale-flecked Morethia (*Morethia obscura*), Death Adder (*Acanthophis antarcticus*) and the Southern Blind Snake (*Ramphotyphlops australis*).

*Granite outcrop,
Canning Dam*

These rock formations in the Darling Range support many reptiles.

A sample of species found in this habitat are the Western Granite Worm Lizard (*Aprasia pulchella*), Ornate Crevice Dragon (*Ctenophorus ornatus*), Red-legged Ctenotus (*Ctenotus labillardieri*), South-western Crevice Egernia (*Egernia napoleonis*), Southern Carpet Python (*Morelia spilota*), Gould's Snake (*Rhinoplocephalus gouldii*) and the Black-backed Snake (*R. nigriceps*).

Wetland, Lake Joondalup

Freshwater lakes, streams, swamps and catchment areas are found in the Darling Range and on the Swan Coastal Plain. With their fringing vegetation, mainly consisting of bulrush and paperbark (*Melaleuca* spp.) as well as Flooded Gum (*Eucalyptus rudis*) on the plain, they support a number of frogs and reptiles.

A sample of species found in this habitat are all four local froglets (*Crinia* spp.), Guenther's Toadlet (*Pseudophryne guentheri*), Slender Tree Frog (*Litoria adelaidensis*), Western Green Tree Frog (*L. moorei*), Oblong Turtle (*Chelodina oblonga*), South-western Cool Skink (*Bassiana trilineata*), Western Glossy Swamp Egernia (*Egernia luctuosa*) and the Tiger Snake (*Notechis scutatus*).

Human-made grassland, Canning Vale

The clearing of native vegetation usually allows the dense growth of introduced grasses. Several reptiles appear to thrive in this new habitat.

A sample of the reptile species found here are the Burton's Legless Lizard (*Lialis burtonis*), West-coast Four-toed Lerista (*Lerista elegans*), Common Dwarf Skink (*Menetia greyii*), Bobtail (*Tiliqua rugosa*) and the Dugite (*Pseudonaja affinis*).

Rottnest, on the other hand, is a tourist resort and has a long history of human disturbance, including farming and the introduction of exotic weeds and pests. In spite of this it has a greater diversity of herpetofauna than Garden Island. The island Bobtail (*Tiliqua rugosa konowi*) and Dugite (*Pseudonaja affinis exilis*), like their mainland counterparts, seem to have adapted well to the changes.

Conservation

Urban development places considerable pressure on all wildlife in a city environment, including frogs and reptiles. As the urban sprawl expands, much of the natural bushland and wetland in and around Perth is changed for all time. Only the most adaptable and less habitat-specific animals persist in urban areas. A remnant pocket of bushland in an inner city suburb will support many frogs and reptiles. These pockets of bushland are very important but, because the surrounding areas are cleared for roads and buildings, they are slowly being degraded. In addition to this slow habitat destruction there is the pressure of introduced predators such as cats, dogs and kookaburras.

Roughly six per cent of Western Australia is set aside for national parks and nature reserves, which compares favourably with the four per cent set aside nationally. This land is managed by the Department of Conservation and Land Management. Protecting these areas and the habitats they support is a prerequisite to conserving the wildlife living there, but it is not sufficient. Protection is far from synonymous with conservation. Once an area has been protected, then for conservation to be sustainable, we must consider a complete range of human

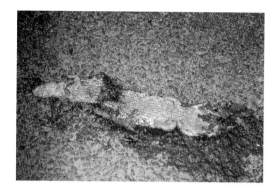

Roadkill Bobtail

Remains of an Oblong Turtle, probably killed by a fox

activities and their impact on the indefinite survival or natural demise of the life forms existing there. Ongoing studies are needed to gauge this impact, identify the life forms and determine their biology and ecology. Without this information, we cannot effectively conserve our wildlife.

Much of the information that is vital in conserving frogs and reptiles has been, and will continue to be, collected by those with an amateur interest. Therefore there must always be a positive relationship between the relevant wildlife authorities and genuine amateur naturalists.

One of the aims of the Western Australian Wildlife Conservation Act 1950–1979 is to protect wildlife against poachers and careless amateur collectors. Unfortunately, it may also hamper the studies of genuine amateur naturalists because it incorporates such broad definitions that it could be used by uninformed or unscrupulous bureaucrats to intimidate whomever they like. The definition in this Act of 'to take', in relation to fauna, is so broad that all residents of and visitors to Western Australia, have at some time committed an offence under the Act. We believe that the current legislation and wildlife policies do little to assist conservation.

Western Australian and Commonwealth legislation tends to place too much conservation value on the protection of individual animals, rather than on conservation of habitat. The substantial resources directed towards the enforcement of this protection would be far better used for the improved management of reserves.

The main factor contributing to the demise of local reptiles and frogs is the removal of their habitat. Development is necessary, but let's hope that in future more land-use decisions will be made with conservation values in mind. Other factors that endanger the survival of our fauna are bushfires, overgrazing and introduced predators.

Special protection of 'threatened' species is provided for under the Wildlife Conservation Act, at the discretion of the Minister for the Environment. It is our opinion that three reptiles found in this area are in need of such protection. One is already on the gazetted list, the Western Swamp Turtle (*Pseudemydura umbrina*). The others, the Perth Lined Lerista (*Lerista lineata*)

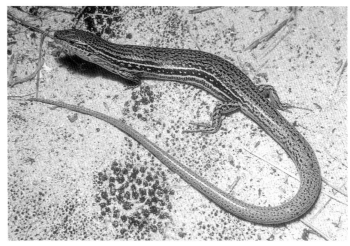

Ctenotus lancelini,
Lancelin Island

and the Black-striped Snake (*Neelaps calonotus*) were removed from the list in 1990. We believe both should be reinstated to the list because of their limited distribution. They are found only within the area covered by this book. In the short term they seem safe because there are nature reserves in this area. But these reserves are undergoing slow degradation because of human activities.

Two threatened species occur just outside the Perth region. The Lancelin Island Skink (*Ctenotus lancelini*) is restricted to the nine hectares of Lancelin Island and small area of coast near Lancelin town site. Recent surveys on the island have shown its numbers to be declining (Browne-Cooper and Maryan, 1992). At the time of writing, the Department of Conservation and Land Management has implemented a Lancelin Island Skink recovery program, involving both professional and amateur naturalists.

Aspidites ramsayi,
Watheroo

The Woma or Sand Python (*Aspidites ramsayi*), particularly the population in the south-west of the State, has been severely affected by agricultural development and introduced predators. The few individuals observed in recent years, with one exception from Badgingarra, have been old adults. More resources are needed to determine its true status within reserves of the south-west. We are pleased to note a move within the Department of Conservation and Land Management to monitor python populations occurring in the south-west of the State.

FROGS
(Class Amphibia, order Salientia)

OST of us are familiar with these fascinating animals. Their bulging eyes, hopping gait and distinctive calls are the stuff of fairy tales and superstitions. In recent times, many biologists have come to consider them the environmental indicators of the future. This is because they are usually closely associated with water. It is believed that their decline may be related to the deteriorating quality of our water through pollution and/or to changes in climate that affect humidity.

The term 'amphibian' is derived from the Greek words *amphi* meaning 'both' and *bios* meaning 'life'. This refers to their two lives: the first as a larva or tadpole and the second as an adult or frog. There are about 3500 species of amphibian worldwide (including toads). Of these, about 200 are found in Australia and 80 or so in Western Australia. Five families are represented on this continent, one of which has been introduced. The well-known Cane Toad (also called Marine Toad), *Bufo marinus*, of the family Bufonidae was introduced from Hawaii in 1935, to combat beetles in the northern Queensland sugarcane fields. This is the only true toad found in Australia. The remaining four families are the true frogs (Ranidae), narrow-mouthed frogs (Microhylidae), water-holding and tree frogs (Hylidae) and ground frogs (Myobatrachidae). The last two are the only families found in Western Australia.

Like many night-active (nocturnal) animals, frogs usually have large eyes. During the day the pupil of the eye is reduced to a vertical or horizontal ellipse. The ear, or tympanum, is located just behind the eye. This may be clearly visible, as is often the case in smooth-skinned frogs, or hidden. Where it can be seen it is a circular disc, which in some species is almost as large as the eye. As would be expected in an animal that can make a lot of noise, frogs have very well-developed hearing.

The adaptations that different species have made to particular environments are best illustrated by the size and shape of their limbs. The climbers, such as the *Litoria* species, have very long legs with expanded terminal discs on each digit. These discs aid climbing by allowing them to grip smooth surfaces. The other extreme is found in the burrowers. These have short strong legs that are used as shovels. On the outside of each foot there is a bony tubercle or 'spade', which allows them to excavate a burrow when shuffling backwards. Although most burrowing species dig in this way, there are two Western Australian endemic (restricted to WA) species that excavate while moving forwards: the Sandhill Frog (*Arenophryne rotunda*) and the Turtle Frog (*Myobatrachus gouldii*).

The skin of frogs is moist but not water-tight. A frog will dehydrate quickly if carried around in your pocket or caught out in the open when the sun rises. For this reason, the greatest variety of species in Western Australia occurs in the higher rainfall areas; the south-west and the Kimberley. Even so, some species have adapted to very dry arid regions. These will burrow deep into the soil and produce a secretion from their skin that forms a membranous cocoon to protect them against dehydration until it rains. A frog's skin may be smooth or covered in wart-like tubercles. (By the way, you will not get warts from handling frogs!)

During the breeding season male frogs call, both to attract a mate and to stake their claim to a patch of territory. Generally, the small frogs emit high-pitched calls and the large frogs low-pitched calls. The sound is produced by the movement of air across the vocal cords. Before calling the male frog inhales, filling the lungs. Then, closing the mouth and nostrils, he transfers the air forwards, inflating the vocal pouch or sac. As the air moves back to the lungs and forwards again it passes across the vocal cords, causing a sound. Often the common name of a frog is derived from its call, for example, the Moaning Frog (*Heleioporus eyrei*) and the Pobblebonk or Western Banjo Frog (*Limnodynastes dorsalis*).

All frogs lay eggs. These consist of a spherical ovum covered with clear jelly and enclosed in an outer capsule. A clump of eggs is generally referred to as 'spawn' or 'egg mass'. The clumps laid by different species vary considerably. The eggs may be laid singly, in small clumps, as a foam nest or in string-like connected chains. They may be laid on or beneath the surface of water, down a flooded burrow, beneath vegetation or in soil. The eggs develop into tadpoles, which emerge, grow and eventually undergo physical changes (metamorphosis) to become frogs. The tadpoles themselves are almost as variable as the adult frogs. Not all frogs have aquatic tadpoles. In the Turtle

Frog (*Myobatrachus gouldii*), metamorphosis occurs within the egg, resulting in the hatching of a fully developed small frog.

Frogs have voracious appetites, feeding on a variety of small invertebrates and vertebrates. If it moves and is small enough to swallow, a frog will eat it. While walking in swampland, Brian Bush was attracted by the squeals of a frog. He found a Western Green Tree Frog (*Litoria moorei*) that had a Slender Tree Frog (*Litoria adelaidensis*) by the back leg and was eating it. Needless to say, the meal was not impressed, and was letting the world know with its squeals.

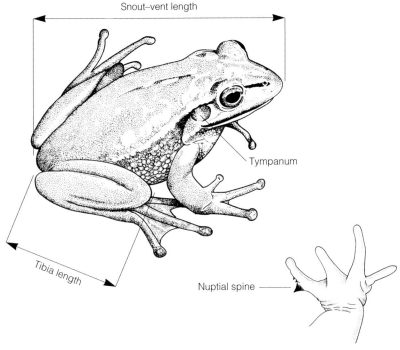

Diagnostic characteristics of frogs

Key to the frogs

1	Tips of fingers with discs	2
	Tips of fingers without discs	3

2	Continuous white stripe along upper lip and side of body	**Slender tree frog** (*Litoria adelaidensis*)
	No white stripe present	**Western green tree frog** (*Litoria moorei*)

3	Large oval gland on upper surface of calf; white vertebral stripe usually present	**Western banjo frog** or **Pobblebonk** (*Limnodynastes dorsalis*)
	No gland on upper surface of thigh, if vertebral stripe present then reddish	4

4	Back and sides with numerous large whitish spots	**Spotted burrowing frog** (*Heleioporus albopunctatus*)
	Back lacks numerous whitish spots, if spots present then restricted to flanks	5

5	Back uniform chocolate to coppery brown	6
	Back marbled or blotched grey, brown or yellow	7

6	Yellow spots and blotches on flanks	**Yellow-flanked burrowing frog** (*Heleioporus barycragus*)
	No yellow spots or blotches on flanks	**Chocolate burrowing frog** (*Heleioporus inornatus*)

7	Back rough. Reddish vertebral stripe usually present	**Humming frog** (*Neobatrachus pelobatoides*)
	Back smooth, no reddish vertebral stripe	8

Key to the frogs (continued)

8	Flanks with fine yellow spots. Call a single low moan	**Moaning frog** (*Heleioporus eyrei*)	
	Flanks lack fine yellow spots, if spots present then white. Call not a single low moan	9	

9	Flanks with fine white spots	**Marbled burrowing frog** (*Heleioporus psammophilus*)
	Flanks without fine white spots	10

10	Red patches in groin and on thighs, eyelids yellowish (or reddish)	**Red-thighed froglet** (*Crinia georgiana*)
	No red patches in groin or on thigh, eyelids not yellowish	11

11	Belly unmarked, body turtle-like, eyes small	**Turtle frog** (*Myobatrachus gouldii*)
	Belly marked with grey, black or brown, body typically frog-like	12

12	Able to hop but usually walks	**Guenther's toadlet** (*Pseudophryne guentheri*)
	Able to walk but usually hops	13

13	Belly smooth	**Green-bellied froglet** (*Geocrinia leai*)
	Belly granular	14

14	Call a squelch. Restricted to Swan Coastal Plain	**Sandplain froglet** (*Crinia insignifera*)
	Call a bleat or rattle, does not resemble a squelch. Found both on the Swan Coastal Plain and in the Darling Range	15

15	Call a bleat, throat in males not completely black	**Granite froglet** (*Crinia pseudinsignifera*)
	Call a rattle, throat in males completely black	**Glauert's froglet** (*Crinia glauerti*)

Ground frogs
(Family Myobatrachidae)

UNTIL recently, members of this family were placed in the southern frog family, Leptodactylidae. This family has now been defined as being restricted to the Americas and southern Africa. The Myobatrachidae is a New Guinean-Australian regionally endemic family.

All local members of this family are small to medium-sized ground-living frogs. They can be distinguished easily from Australian members of the family Hylidae, which contains the tree frogs and water-holding frogs. Ground frogs do not have finger and toe pads (or if they do, they are poorly developed), neither do they have the extra bone (called the intercalary cartilage) found in the fingers and toes of the tree frogs. The water-holding frogs (*Cyclorana* species) are ground-living frogs, but are not represented in the Perth area.

Members of the Myobatrachidae family display considerable diversity in morphology, life-cycle and ecology. Most are terrestrial or burrowing, but some are aquatic. None are primarily tree-living. Egg and tadpole development ranges from fully aquatic to fully terrestrial. In the fully terrestrial species there is no free-living tadpole stage; a fully developed baby frog emerges from the egg.

This family is well represented in the Perth region, with eight genera and 14 species.

Red-thighed froglet
Crinia georgiana Tschudi 1838

Genus *Crinia*
Tschudi 1838
Small depressed frogs. Vomerine teeth present or absent. Eardrums visible or absent. Limbs short, fingers and toes long and unwebbed, toes without fringe to slightly fringed. Belly granular. Eggs pigmented and laid singly in water. Tadpole stage in water or waterlogged soil.

Dwellingup

HABITAT	Patchily distributed on the Swan Coastal Plain, usually in and around ephemeral swamps. More abundant in the Darling Range along streams and gullies and in moist situations such as under granite rocks and logs.
DESCRIPTION	A small somewhat flattened frog. Considerable variation in back pattern and colour. All have a bright red patch in the groin and on the thighs, and yellow or reddish upper eyelids. Adult SVL: ♂ to 32 mm; ♀ to 36 mm.
MATING CALL & BREEDING BIOLOGY	A short loud call described as a 'quack … quack … quack' (Main, 1965). Breeding occurs on cold nights from July to October. The eggs are laid in any available shallow pool of water. Metamorphosis takes 35–45 days.

Glauert's froglet
Crinia glauerti Loveridge 1933

Melaleuca Park

HABITAT — Found both on the Swan Coastal Plain and in the Darling Range, around areas of permanent moisture. When the surface moisture dries during summer, it descends to the damp soil beneath.

DESCRIPTION — Highly variable in back pattern and colour, from monotonal brown or black to darker and lighter patches, zones and lines. Often some indication of reddish striations along thigh. Dorsal skin with or without numerous tubercles and ridges. Belly white or grey, often with black patches and variegations. Adult SVL: ♂ to 22 mm; ♀ to 24 mm.

MATING CALL & BREEDING BIOLOGY — Call described as a long hollow rattle, not unlike 'a pea in a can' (Main, 1965). It is an opportunistic breeder, breeding following rain at any time except in mid-summer. The eggs are laid in still pools of water, where they sink to the bottom.

Sandplain froglet
Crinia insignifera Moore 1954

Melaleuca Park

HABITAT — Found on the Swan Coastal Plain, including Rottnest Island, in ephemeral swamps, low-lying areas subject to winter flooding and permanent watercourses. Absent from the Darling Range, where the Granite Froglet (*Crinia pseudinsignifera*) fills a similar niche.

DESCRIPTION — Colour highly variable between individuals, from monotonal grey, brown or black to a complex series of darker and lighter longitudinal patches, lines and zones. Often there is a dark triangular-shaped patch on the head between the eyes. Dorsal skin with or without numerous tubercles and ridges. Adult SVL: ♂ to 23 mm; ♀ to 30 mm.

MATING CALL & BREEDING BIOLOGY — Call described as a 'squelch', not unlike dragging wet fingers across a balloon. It is a winter breeder, laying eggs singly in still pools, where they sink to the bottom. Metamorphosis takes up to 150 days (Main, 1957 & 1965). It is known to hybridise with *Crinia pseudinsignifera*.

Granite froglet
Crinia pseudinsignifera Main 1957

Parkerville

HABITAT	Locally restricted to the Darling Range. Found in association with granite, especially around the base of outcrops where run-off water collects.
DESCRIPTION	Colour highly variable between individuals, from monotonal grey, brown or black to a complex series of darker and lighter patches, lines and zones. Often there is a dark triangular-shaped patch on the head between the eyes. Dorsal skin with or without numerous tubercles and ridges. Adult SVL: ♂ to 24 mm; ♀ to 25 mm.
MATING CALL & BREEDING BIOLOGY	Call described as a high-pitched wavering 'baa ... baa ... baa' (Main, 1957). Locally, it is a winter breeder, although at least one population elsewhere breeds in summer. Eggs are laid in pools and adjacent waterlogged soil. Metamorphosis takes 80–130 days (Main, 1965).

Green-bellied froglet
Geocrina leai (Fletcher 1898)

Genus *Geocrinia*
Blake 1973

Head and body broad and marginally depressed. Vomerine teeth usually present but often inconspicuous. Eardrums inconspicuous or hidden. Limbs short, fingers and toes unwebbed, toes with or without terminal discs. Belly smooth to granular. Eggs pigmented and laid in a clump. Tadpole stage found in water.

G. Harold   Walpole

HABITAT	Locally, restricted to the Darling Range, close to swamps and streams.
DESCRIPTION	A small, somewhat flattened, frog. Dorsal skin may be smooth but more often covered with numerous tubercles. Brown to yellowish with a broad, dark (occasionally white-edged) mid-dorsal region or series of blotches. Belly greenish. Adult SVL: ♂ to 21mm; ♀ to 26mm.
MATING CALL & BREEDING BIOLOGY	Call described as a 'chic ... chic ... chic ... chic' or 'tik ... tik ... tik ...tik'. A winter breeder that lays a clump of eggs attached to vegetation above or adjacent to water. Upon hatching the tadpoles enter the water. Metamorphosis takes more than 120 days (Main, 1965).

Spotted burrowing frog
Heleioporus albopunctatus Gray 1841

Genus *Heleioporus*
Gray 1841

Moderately large and robust. Head short. Pupils vertical. Vomerine teeth prominent. Eardrums hidden to prominent. Parotoid glands slightly raised. Limbs short, fingers unwebbed, first finger longer than second, toes partly webbed, shovel-shaped inner metatarsal tubercle. Eggs large, pigmented or unpigmented and laid out of water in a foam nest. Tadpole stage found in water.

Mundaring

HABITAT   Locally restricted to the Darling Range. Its burrows have been found in the banks and under stones on the bed of shallow ephemeral watercourses, in swamps or at the vertical edges of claypans.

DESCRIPTION   A large and very robust purplish to chocolate-brown frog with numerous evenly spaced pale spots on the back, sides and limbs. Tympanum hidden. Adult males lack black spine on first or second fingers. Adult SVL: ♂ to 77 mm; ♀ to 85 mm.

MATING CALL & BREEDING BIOLOGY   A short high-pitched 'coo' repeated at a rate of slightly more than one per second. Eggs are laid in a burrow in the form of a foam nest, and take 10–40 days to hatch (Lee, 1967). It is believed to hybridise with *Heleioporus eyrei* in the Darling Range. Hybrids have a call resembling that made by a choking cat (K. Aplin, personal communication).

Yellow-flanked burrowing frog
Heleioporus barycragus Lee 1967

Gidgegannup

HABITAT
Generally restricted to the Darling Range, north to about Bullsbrook. It may be found on clay or granite adjacent to turbulent winter-flowing watercourses (Main, 1965).

DESCRIPTION
A large, robust greyish to chocolate-brown frog with numerous bright yellow wart-like spots and blotches on flanks. Some individuals have a black tubercle at the centre of each of these pale spots. Tympanum prominent. Adult males have a black thorn-like nuptial spine on first finger and other smaller spines on the first and second fingers. Adult SVL: ♂ to 83 mm; ♀ to 86 mm.

MATING CALL & BREEDING BIOLOGY
Call described as a 'low-pitched, owl-like hoot', slowly repeated. Eggs laid in a burrow excavated by the male in a nearly vertical bank (Lee, 1967).

Moaning frog
Heleioporus eyrei (Gray 1845)

Neerabup

HABITAT	Very common on the Swan Coastal Plain, where it is often unearthed in suburban gardens. Less common in the Darling Range. Also occurs on Rottnest Island. Usually on sandy soils near swamps and streams.
DESCRIPTION	Brown to dark grey, diffused with irregularly defined yellowish patches. Flanks finely peppered with white spots. Ear indistinct to prominent. Adult males lack black nuptial spines on first or second fingers. Adult SVL: ♂ to 66mm; ♀ to 63mm.
MATING CALL & BREEDING BIOLOGY	Call a long and rising low moan, slowly repeated. The call often causes sleepless nights for people living near wetlands. Eggs are laid in an oblique burrow dug in sand where the land is horizontal (Lee, 1967).

Chocolate burrowing frog
Heleioporus inornatus Lee and Main 1954

Chidlow

HABITAT	Restricted to the Darling Range in the vicinity of: "sandy, acid peat bogs" (Main, 1965).
DESCRIPTION	Uniform copper-brown with or without a mottling of white, yellow or grey. Tympanum indistinct. Adult males with one or two small black nuptial spines on first finger. Adult SVL: ♂ to 64 mm; ♀ to 73 mm.
MATING CALL & BREEDING BIOLOGY	Call described as a 'woop-woop' repeated frequently. Eggs are pale yellow in colour and laid in a burrow dug into the side of a sloping or vertical bank (Lee, 1967).

Marbled burrowing frog
Heleioporus psammophilus Lee and Main 1954

Parkerville

HABITAT
Found on the Swan Coastal Plain and the western edge of the Darling Range, where the substrate is angular fine-grained sands and sandy clays (Main, 1965).

DESCRIPTION
Purplish brown to dark brown with indistinct to bold white or pale grey patches on back. If the dorsal pattern is indistinct, then flanks have small white spots. If the dorsal pattern is bold then these spots are less discernible. Tympanum indistinct to prominent. Adult males with or without black nuptial spine on first finger. Adult SVL: ♂ to 62 mm; ♀ to 60 mm.

MATING CALL & BREEDING BIOLOGY
Main (1965) described the call as a short high-pitched 'put-put-put-put', similar to a single-cylinder motor such as a small outboard or lighting plant. Breeding biology believed to be similar to that of the Moaning Frog (*Heleioporus eyrei*).

Western banjo frog or Pobblebonk *Limnodynastes dorsalis* (Gray 1841)

Genus *Limnodynastes*
Fitzinger 1843
Body moderately large and robust. Head large with slightly flattened snout. Pupils vertical. Eardrums prominent. Vomerine teeth behind internal opening of nostrils. Limbs muscular and short to moderately long. Fingers unwebbed, toes webbed or unwebbed. Eggs pigmented and laid in foam nest on water. Tadpole stage found in water.

Waterman

HABITAT	This frog may be found throughout the Perth area. It occupies the vegetation adjacent to water during the wet winter months, moving away from water to seek concealment in a burrow during summer.
DESCRIPTION	A large reddish brown to golden frog with bold black patches over back and hindlimbs. A black bar from eye to above forelimb, bordered below by a white bar terminating posteriorly at a white to yellow supralabial gland. Usually a narrow pale mid-dorsal stripe from nose to between hindlimbs. Groin bright red. Toes very slightly webbed; shovel-shaped inner metatarsal tubercle. Adult SVL: ♂ to 64 mm; ♀ to 73 mm.
MATING CALL & BREEDING BIOLOGY	Call is a single deep explosive 'bonk', similar to that made by a banjo string. Breeding occurs in late autumn, winter and spring. Males may call from the shelter of vegetation or from the edges of shallow pools. Male–male combat has been recorded at night in late May (Bush, 1984a). In this instance, the combatants both looked and behaved like sumo wrestlers. Two males had several wrestling bouts spaced every two or three minutes over a 25-minute period. The end came when both combatants moved away from the calling site: a small shallow pool of water. Eggs are laid in the form of a foam nest on the surface of still or slowly flowing water (Main, 1965).

Turtle frog
Myobatrachus gouldii (Gray 1841)

Genus *Myobatrachus*
Schlegel 1850
A genus containing a single species (monotypic). This species has a bulbous body, small head and short stubby limbs. Eyes very small and inconspicuous. Eardrum hidden. Fingers and toes short and unwebbed. Eggs large and unpigmented. Tadpole stage out of water.

Ellenbrook

HABITAT	Locally, found on the Swan Coastal Plain. Absent from the Darling Range. Occurs most commonly on sand, where it has been uncovered beneath large logs and raked out of spoil-heaps. It seems to feed entirely on termites and is often associated with colonies of termites in mounds, stumps and infested wood.
DESCRIPTION	Pale yellowish brown, grey or dark brown. Adult SVL: ♂ to 42 mm; ♀ to 50 mm. For further description, see notes under genus.
MATING CALL & BREEDING BIOLOGY	Call described as an abrupt, deep 'croak'. (Here, we have a most unlikely-looking frog that emits the traditional frog croak!) Eggs have been found at depths of 1–1.2 metres in moist sand (Roberts, 1981). These are large, 5–7 mm in diameter, and are laid in clumps of up to 40 individual eggs. Development to frog stage occurs within the egg capsule. There is no free-swimming tadpole.

Humming frog
Neobatrachus pelobatoides (Werner 1914)

Genus *Neobatrachus*
Peters 1863

Head and body robust. Pupils vertical. Vomerine teeth present. Eardrums indistinct to prominent. Limbs short. Fingers unwebbed, first longer than second. Toes partly webbed, black or unpigmented, large shovel-shaped inner metatarsal tubercle. Eggs pigmented and laid in a long string. Tadpole stage in water.

Chidlow

HABITAT	Locally most common on Darling Range, south to Jarrahdale. Found in association with clays or loams.
DESCRIPTION	A stout, medium-sized, yellow to blackish brown or greenish frog, often irregularly covered with paler or darker patches. Usually having a fine red or yellow vertebral stripe over the head and body. Tympanum prominent. Adult SVL: ♂ to 43 mm; ♀ to 44 mm.
MATING CALL & BREEDING BIOLOGY	Call is a low-pitched trill or hum, only audible over a short distance (Main, 1965). Breeding takes place from May to July in claypans temporarily filled with water. See notes on genus concerning eggs.

Guenther's toadlet
Pseudophryne guentheri Boulenger 1882

Genus *Pseudophryne*
Fitzinger 1843
Body small and squat. Head small. Pupils horizontal. Vomerine teeth absent. Eardrums absent (present in one species from the deep south-west of Western Australia). Limbs short. Fingers unwebbed, first finger shorter than second. Toes unwebbed, inner metatarsal tubercle small or large. Eggs large, unpigmented and laid out of water (except possibly one species from the arid north-west of the State). Early tadpole stage out of water, remaining stages in water.

Melaleuca Park

HABITAT	Found both on the Swan Coastal Plain and in the Darling Range, in moist areas beneath rocks, logs, deadfall vegetation and rubbish.
DESCRIPTION	A small robust frog with irregularly mottled grey, brown and orange back. Belly white with large black patches. Metatarsal tubercles well developed. Although it can hop, it usually walks. Adult SVL: ♂ to 30 mm; ♀ to 33 mm.
MATING CALL & BREEDING BIOLOGY	Call described as a short harsh grating 'squelch'. Males call from burrows in the soil (Main, 1965). Eggs laid separately in a burrow, following heavy rains from late summer to winter. The tadpoles hatch at an advanced stage of development, when the burrow floods.

Tree frogs and water-holding frogs
(Family Hylidae)

A large family of frogs found throughout most of the temperate and tropical regions of the world. The Australian members of this family may eventually be recognised as an Australasian regionally endemic family, the Pelodryadidae.

These are small to large arboreal, terrestrial and burrowing frogs that are distinguished from members of the ground frog family (Myobatrachidae) on the basis of the following characteristics: fingers and toe pads poorly developed to well-developed, and an extra bone (intercalary cartilage) present between the last two bones of each digit in most members. They display considerable diversity in morphology and ecology. Many are efficient climbers, with long legs and large adhesive toe pads, whereas others are unable to climb. The eggs of all Australian species are believed to be laid in or on water, and develop into free-swimming aquatic tadpoles.

Locally, only one genus and two species of tree frog are represented. The water-holding frogs (*Cyclorana* species) are ground frogs, but are not represented in the Perth area.

Slender tree frog
Litoria adelaidensis (Gray 1841)

Genus *Litoria*
Tschudi 1838

Body small to large, slender to robust. Pupils horizontal. Vomerine teeth present or absent. Eardrums at least partly visible. Limbs short to long. Fingers and toes with terminal discs, fingers webbed or unwebbed, toes webbed. Eggs pigmented and laid singly, in small groups or as a large cluster on or beneath the surface of the water. Tadpole stage found in water.

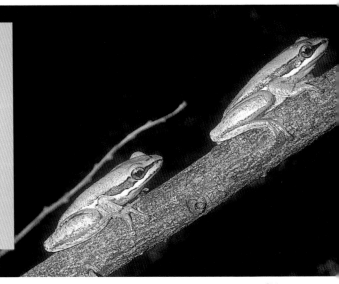

Gidgegannup

HABITAT	Found on the Coastal Plain and western edge of the Darling Range inland to about Mundaring. Usually in the vicinity of dense vegetation adjacent to still or slowly moving water. Often observed basking during the day on bulrushes and sedges, escaping to the base of the plant when disturbed.
DESCRIPTION	A slender green or brown frog with elongated head. Some individuals may be brown with large green patches or vice versa. Usually there is a broad dark stripe along the side of the head and body, bordered below by an equally broad white stripe. Bright red, orange or yellow spots or blotches on the backs of thighs. Tympanum prominent. Back smooth, without tubercles. Fingers unwebbed, toes webbed. Adult SVL: ♂ to 45 mm; ♀ to 47 mm
MATING CALL & BREEDING BIOLOGY	Call described as a 'harsh grating screech'. Breeding occurs in spring. Males call from water or near water, from ground or vegetation above ground (Main, 1965). Eggs laid as an irregular clump attached to vegetation on the surface of water.

Western green tree frog
Litoria moorei (Copland 1957)

Lynwood

HABITAT — Found both on the Swan Coastal Plain and in the Darling Range near permanent water, including backyard ponds and ephemeral swamps. Occurs on Rottnest Island. Often discovered beneath bark on standing or fallen trees, beneath exfoliated granite and beneath rubbish.

DESCRIPTION — A large, robust green or brown frog with triangular-shaped head. Typically, the ground colour is interspersed with clearly defined regular dark patches, that is, pale brown with dark green or dark brown patches. There is a pale vertebral stripe and a dark stripe from the eye passing through the ear to the flank, midway between the fore and hind limbs. This stripe may extend forward from the eye to the nostril, and is often bordered above with white. Tympanum prominent. Back with a scattering of distinct tubercles. Fingers unwebbed, toes webbed. Adult SVL: ♂ to 64 mm; ♀ to 74 mm.

MATING CALL & BREEDING BIOLOGY — Call described as a long low growl broken into several parts, not unlike a motorbike changing gears. Eggs are laid in a floating clump attached to surface vegetation (Main, 1965).

SEA TURTLES
(Order Testudines, family Cheloniidae and Dermochelyidae)

THESE creatures are generally confined to tropical and warm temperate waters. They are characterised by well-developed paddle-shaped limbs without obvious ankle joints. Sea turtles can swim at considerable speed. Their nostrils are positioned on top of the snout, as with other aquatic and marine reptiles. The shell, which is often covered in algae and barnacles, is very strong and consists of bony plates (modified scales). Mating occurs at sea and females come ashore at night on a rising tide to lay their eggs. These are laid in a hole excavated on a sandy beach. Many of the eggs and hatchlings fall prey to predators. Most of the species inhabiting Australian waters are known to breed at specific sites along our northern coastline.

Although some species of sea turtle occur naturally in the Perth region, they are not often seen in our waters. None are known to breed this far south. For this reason, we will not discuss them in detail. The four species observed locally are the Green Turtle (*Chelonia mydas*), Loggerhead (*Caretta caretta*), Hawksbill (*Eretmochelys imbricata*) and the Luth or Leathery Turtle (*Dermochelys coriacea*). The Leathery Turtle is the world's largest, attaining 2.4 metres in length. It is placed in a family on its own (Dermochelyidae). It has a unique shell made up of very small plates embedded in a leathery skin, rather than a fused bony shield.

The Hawksbill Turtle and the Green Turtle have been recorded as having poisonous flesh at times. Both species are believed to have caused death in humans after being eaten. However, no deaths have been attributed to turtle poisoning in Australia for the past 50 years.

Chelonia mydas, *Broome*

H. Ehmann

Key to sea turtles

1	Limbs with claws (these are indistinct spurs on the flippers)	2
	Limbs without claws	**Luth** or **Leathery turtle** (*Dermochelys coriacea*)

2	Snout without pronounced beak in profile, dorsal scales of carapace not overlapping	3
	Snout with pronounced beak in profile, dorsal scales of carapace strongly overlapping	**Hawksbill turtle** (*Eretmochelys imbricata*)

3	Colour greenish, single pair of prefrontals	**Green turtle** (*Chelonia mydas*)
	Colour brownish, two pairs of prefrontals	**Loggerhead turtle** (*Caretta caretta*)

FRESHWATER TURTLES
(Family Chelidae)

A FAMILY of aquatic and semi-aquatic reptiles found also in New Guinea and South America. They belong to a group described as pleurodirous or side-necked turtles. This refers to the way they retract the head by folding the neck into a horizontal groove under the front edge of the upper shell. This differs from other freshwater turtles, which fold the head into an S-shape to withdraw it. The limbs can also be withdrawn to some extent. All have distinct ankle joints, webbed feet and four or five claws.

Male turtles have much longer tails than females. Worrell (1963) describes courtship as '… an uncomplicated affair; the male circles the female and occasionally strokes her face gently with a fore-foot. Mating takes place under the water'. The eggs, when laid, are white and brittle-shelled.

We refer to the members of this family as turtles because of their webbed feet and aquatic behaviour. The term 'tortoise' is used for the dome-shelled, predominantly land-based tortoises, such as those found on the Galapagos Islands. Sixteen species occur in Australia; two are found locally.

Long-necked or Oblong turtle
Chelodina oblonga Gray 1841

Genus *Chelodina*
Fitzinger 1826

Exceptionally long-necked turtles, with the extended neck being as long or longer than the shell. Members are distinguished from other freshwater turtles by having only four clawed toes on the front feet, instead of five.

G. Kuchling — Lake Joondalup

HABITAT
Common in permanent freshwater and seasonal swamps throughout the Perth region.

DESCRIPTION
Head, neck and carapace (top shell) pale to dark brown or black, often with darker flecks and variegations. Carapace usually covered with algae, which makes them difficult to see. Plastron (bottom shell) whitish. The name 'Oblong' is derived from the shell shape. When viewed from above, the adult carapace is distinctly elliptic. Shell length to 40 cm.

GENERAL
Spends most of the time in water, remaining submerged for extended periods. When surfacing to breath, only the nostrils and tip of snout may be visible briefly before it dives again. This largely aquatic lifestyle is interrupted by short basking spells on partly submerged logs or rocks, quickly plopping into the water if disturbed. It has been observed migrating between bodies of water. When females are ready to lay eggs, they may travel considerable distances over land to find suitable laying sites. These are in positions that will not be inundated with water during winter rains. During these travels, many are killed on roads.

Clutches of up to 25 eggs are laid from October to February, and take many months to hatch. Large numbers of hatchlings are found in August and September, with many perishing in grass that is too long to allow

Herdsman Lake

them to get to the water, or being trapped after falling into stormwater drains. Rescued hatchlings, if kept for rehabilitation, often do not feed for several weeks. Eventually, when hungry, they will thrive on garden worms cut into short lengths. The food, if dropped into the tank close to the baby turtle, will be grabbed at lightning speed.

The Oblong Turtle is carnivorous, feeding on fish, molluscs and crustaceans. There are reports of adult turtles taking ducklings from the surface of the water. No doubt some water-birds in turn feed on the baby turtles. When handled they often leave a foul-smelling odour on the hands. It is produced by glands located at each section of the bridge where the carapace joins the plastron.

Short-necked or Western swamp turtle *Pseudemydura umbrina* Siebenrock 1901

Genus *Pseudemydura*
Siebenrock 1901
A monotypic genus characterised by an extremely short neck, much shorter than the carapace length, and a horny upper part of head (or casque). Forelimbs with five clawed toes.

Ellenbrook

HABITAT	Only known from a few scattered ephemeral (temporary) swamps on the Swan Coastal Plain near Bullsbrook.
DESCRIPTION	Easily identified locally by its very short neck. Head, neck and squarish carapace dark yellowish-brown to black. Plastron white to off-white, almost as wide as carapace. Neck skin tubercular, giving it a wart-like appearance. Shell length to 15 cm.
GENERAL	Discovered in 1839 and described to science in 1901. It was not sighted again until a young amateur herpetologist collected it in 1953 near Bullsbrook. It is estimated that there are only 30 individuals in the wild, and about 70 are held at the Perth Zoo. It is therefore considered by many to be the world's rarest reptile.

When the ephemeral swamps at Ellenbrook Nature Reserve dry out in the summer, this turtle aestivates in deep cracks and fissures in the clay. At Twin Swamps Nature Reserve it was believed to retreat to holes dug in the ground beneath litter such as deadfall vegetation or logs. When the swamps at Ellenbrook fill in June and July it emerges, and may be found

G. Kuchling   A hatching success

swimming in shallow water, feeding on crustaceans, tadpoles and aquatic invertebrates. Up to five eggs are laid in October and November, taking as long as 190 days to hatch.

An ongoing breeding program involving the Perth Zoo, The University of Western Australia and the Department of Conservation and Land Management is aimed at breeding large numbers of these turtles for future release into suitable habitats.

During the preparation of this book several Western Swamp Turtles were found in agricultural dams outside reserves.

LIZARDS AND SNAKES
(Order Squamata)

THE lizards (suborder Lacertilia) and snakes (suborder Serpentes) may be considered the most successful of modern-day reptiles. Together, they are the most species-rich group of terrestrial vertebrates in Australia, with almost 700 species known to occur here. Needless to say, if you have an interest in lizards and snakes, Australia is the place to live. As the group is so diverse and little-known, the number of species recognised is increasing as our knowledge expands.

Although lizards and snakes display many distinctly different characteristics, it is generally believed that snakes evolved from a lizard-like ancestor; possibly a primitive lizard similar to the monitor lizards of today. Perhaps the snakes' limbed ancestors gradually moved to a poorly utilised and therefore less competitive subterranean environment. Large limbs and thick bodies would hinder movement through soil; a more elongated form without protruding limbs would be more successful. Hence today's snakes.

This progression to a slender form and limblessness can be seen in some lizard groups today. Members of the genus *Lerista* display varying degrees of elongation and a reduction in both the size and number of limbs. Those species that forage mostly on the surface have comparatively thick bodies and well-developed limbs, whereas those that spend most of their life burrowing through loose soils are long, slender and almost limbless.

Legless lizards of the family Pygopodidae display similar characteristics, but all members lack fully-developed limbs. The worm lizards (*Aprasia*) are burrowers with only a trace of hindlimbs in the form of scaly flaps, one on each side, and they lack external ear-openings. Members of the genus *Delma* spend little time below ground and have a larger hindlimb flap and obvious ear-opening. The *Pygopus* species have the largest hindlimb flaps and ear-openings and are successful above-ground and semi-arboreal foragers. The Pygopodidae family includes two exceptional species that have unusual characteristics and do not fit the general trend (see *Aclys* and *Pletholax* genera).

What do lizards and snakes have in common?

Being reptiles, both lizards and snakes are cold-blooded (ectothermic). Much of their behaviour associated with controlling body temperature is similar. Those that are active during the day (diurnal) often bask in direct sunlight in the spring and autumn. The sun's warmth raises their temperature to the optimum for metabolism and activity to occur. In mid-summer when air temperatures are high, there is little need for basking. Most activity (or lack of it) at this time is to prevent their temperature from rising to a critical level. Many species may cease moving during the day, foraging at night instead.

The behaviour used to maintain their body temperature near optimum (thermoregulation) is quite complex. A basking reptile may expand, flatten or angle its body towards the sun. A night-time-active (nocturnal) reptile will depress and flatten its body hard against the warmer substrate, be it the earth, a rock or the road. To lower its body temperature, it will lessen the contact between its body and the substrate or move to cooler sites. A lizard may pant with its mouth open.

The Perth region has a temperate mediterranean climate with cool winters. Many of our local reptiles enter a period of torpor or inactivity during the cool months. Some species appear to be less affected by cold than others. The Tiger Snake is cold-tolerant, whereas the Dugite is not. Therefore, you might see a Tiger Snake basking on a sunny winters' day but never a Dugite. A reptile will seek out its winter retreat with the onset of cold weather. This may be the disused burrow of another animal, nooks and crannies among a root system, beneath a rock or log, or any place where the temperature will not fall to zero. When the temperature rises in spring, activity will recommence.

The sense of smell is well developed in most reptiles but particularly so in snakes and those lizards with a forked tongue. A paired sensory organ is located in the roof of the mouth. It consists of two highly odour-sensitive pits that analyse particles conveyed to it on the forked end of the tongue. The constant flicking of the tongue is a notable feature of snakes and some lizards, particularly monitor lizards. They are testing the air to determine whether there is food, a threat, or a mate nearby.

The sense of hearing in snakes and some lizards is not hearing as we understand it. They do not have an external ear-opening. They are not deaf, just lack the ability to detect high frequencies of sound conveyed through the air. They do hear, or more correctly feel, low frequencies, and have a tremendous ability to detect the most subtle of vibrations transmitted through sand, leaf-litter, dried grasses and, to a lesser degree, hard ground.

The eyesight in lizards and snakes is good, especially in detecting movement. Anyone observing a lizard catching insects, or a snake's response to any movement at distances of up to 30 metres will appreciate the acuteness of their eyesight. One has to be careful in feeding captive snakes, for they will respond to their own reflection in the keeper's eyes. The whip snakes (genus *Demansia*) are generally lizard-feeders that catch their prey on the run. The eyesight must be good to maintain visual contact with the darting and dodging lizard. Snakes and even some lizards are not good at seeing stationary objects, but this is offset by the acuteness of the other senses.

All reptiles have dry skin, not wet and slimy skin as many people believe. The outer layer is dead and formed in numerous side-by-side or overlapping scales. Unlike fish, where the scales are attached to the skin, in reptiles the scales are the skin. As the outer layer is dead, it will not grow with the animal and must be shed or sloughed (pronounced 'sluffed') off. In lizards with legs the skin is shed in pieces, but in the snakes and legless lizards it usually comes off in one piece. They crawl out of the old skin turning it inside out as they go. Juveniles may slough as frequently as once every three weeks, whereas old reptiles may only do this 1–3 times a year. A physical feature that most sets the squamates apart from the other reptile groups is the fact that the males have paired sex organs. These are called hemipenes, are located internally at the base of the tail and are used independently.

What are the differences between lizards and snakes?

There are numerous differences between typical limbed lizards and snakes, but there are very few features that differentiate absolutely between the two groups. Many lizards have an elongated form and other snake-like characters. The obvious difference between the typical lizards and snakes is the legs; snakes do not have any. If legs are present then so are shoulder and pelvic bones. Snakes have no shoulder bones and only the most primitive snakes have remnants of pelvic bones.

Eardrums and external ear-openings are present in most lizards but absent in snakes. The lower jaw is one piece in lizards, whereas in typical snakes it is in two halves, divided medially. No Australian snake has the ability to regrow a lost or damaged tail, whereas many lizards can do this. Snakes generally have short tails, rarely amounting to 20 per cent of their total length, whereas long tails are common in lizards. Snakes and legless lizards have internal organs that are modified to fit into a long slender body cavity. The left lung is reduced in size such that it is almost non-existent in some snakes, and the right is lengthened. The liver and kidneys are cigar-shaped, and the kidneys are also staggered, with one located further along the body cavity than the other.

Key to suborders of Squamata

A combination of one or more of the following will identify a squamate as a lizard:

broad fleshy tongue and/or external ear-opening	**Lacertilia**

All snakes have the following:

bifid (forked) tongue and no external ear-opening	**Serpentes**

Lizards

(Order Squamata, suborder Lacertilia)

THE lizards and snakes have been treated as a single group, and you will find an introduction to the group on page 54.

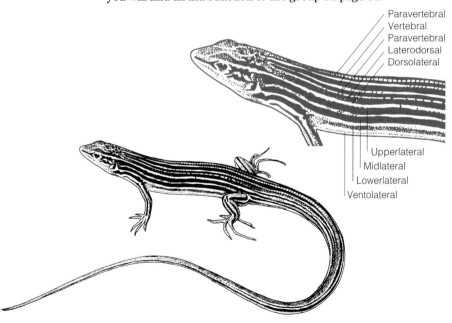

Arrangement of stripes on lizards

Key to lizard families

1. Forelimbs absent, hindlimbs represented by a flap (sometimes barely discernible) — **legless lizard** (Pygopodidae)

   Forelimbs present or absent, hindlimbs present or represented by a stump but never a flap — 2

2. Scales on top of head large and of variable size — **skink** (Scincidae)

   Scales on top of head small and uniform in size — 3

3. Tongue long and forked (bifid), frequently flicked in and out — **goanna** or **monitor lizard** (Varanidae)

   Tongue broad and fleshy, not frequently flicked in and out — 4

4. Eyelid moveable — **dragon lizard** (Agamidae)

   Eyelid immovable — **gecko** (Gekkonidae)

Geckos

(Family Gekkonidae)

THE Gekkonidae are represented in Western Australia by more than 60 species, of which 10 occur naturally in the Perth area. One species, Binoe's Prickly Gecko (*Heteronotia binoei*), that was inadvertently introduced may also be found here.

Geckos appear to be rather fragile, but are well adapted to a wide range of environments, from rain forests to arid deserts. Some can squirt a harmless, mildly odorous fluid from pores along the top of the tail. Most have fragile tails that can be voluntarily dropped or dismember readily when grasped by a predator. When threatened, many can vocalise with an audible bark or buzzing sound.

All Australian species are egg-layers (oviparous), producing one or two eggs in a clutch. Many have multiple clutches, and some breed at any time of year. In some species the shells are brittle, similar to bird eggs, whereas others lay leathery-shelled eggs.

Geckos feed on invertebrates and rarely smaller reptiles. They are almost exclusively nocturnal. Their large eyes with vertically elliptical pupils lack eyelids, but are covered by a protective transparent scale, which is kept clear by wiping with the tongue. These lizards have very specialised fingers and toes. Some are long and bird-like with a fixed claw. One species has no claws, whereas others are able to retract the claws between enlarged finger and toe pads. It is these pads that allow some geckos to run up vertical surfaces as smooth as glass. When examined under a microscope, the pads are seen to be covered with numerous tiny hairs that catch in the smallest imperfections on the surface on which they are climbing.

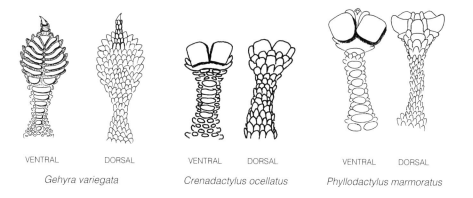

Toe structure in local gecko species

Key to the Gekkonidae genera

1	All or most digits with claws	2
	All digits without claws	*Crenadactylus*

2	All digits clawed	3
	Inner digit on each foot clawless, remainder clawed	*Gehyra*

3	Claws retractile, pair of apical lamellae enlarged	4
	Claws not retractile, no enlarged apical lamellae	7

4	Scales above distal expansion of digit much larger than those above proximal part	*Phyllodactylus*
	Scales above distal expansion of digit not noticeably larger than those above proximal part	5

5	Postanal tubercles large and spur-like, one on each side	*Oedura*
	Postanal tubercles small and in a cluster of two or more	6

6	Dorsal scales heterogeneous, including numerous enlarged tubercle scales. Able to squirt fluid from pores along top of tail	*Strophurus*
	Dorsal scales homogeneous, no enlarged tubercle scales. Unable to squirt fluid from tail	*Diplodactylus* (see p. 66)

7	Tail long and slender, preanal pores present	*Heteronotia*
	Tail broad proximally, tapering abruptly over distal half of length	*Underwoodisaurus*

Clawless gecko
Crenadactylus ocellatus (Gray 1845) subsp. *ocellatus*

Genus *Crenadactylus*
Dixon & Kluge 1964

A monotypic genus confined to Australia. Digits moderately depressed with a distal expansion corresponding to a pair of greatly enlarged squarish subdigital pads (apical lamellae). These are separated from an undivided or divided series of round lamellae by two or three rows of small imbricate scales. Pair of scales above apical lamellae much larger than adjacent scales. All digits without claws. Two preanal pores in males.

Red Hill

HABITAT — Generally found around rock: limestone on the coast, granite and laterite in the Darling Range. On large rock outcrops, it is most common around the fringe where the rock meets the vegetation. Common in the Darling Range, scarce on the Swan Coastal Plain.

DESCRIPTION — Dorsum covered with homogeneous granular scales, each with a weak keel. Ventral scales homogeneous, overlapping, non-granular and smooth. Labials much larger than adjacent scales. A single large cloacal tubercle behind and on each side of the vent, slightly larger in males than females. Ground colour grey to light-brown finely flecked with black. Numerous ocelli may be randomly scattered or aligned in two longitudinal rows, with each ocellus enclosed in a brownish-red paravertebral stripe. A mid-lateral stripe of the same colour from ear to hindlimb. Belly light grey with or without black flecks. Adult SVL: ♂ to 30 mm; ♀ to 35 mm. Tail 65–85% of SVL.

GENERAL — Australia's smallest gecko. Usually uncovered beneath stones, in grass tussocks, amongst deadfall bushes and debris. It lays two (rarely one) soft-shelled eggs, which are 7–8 mm long, 4–5 mm wide and weigh 0.12 g (Bush, 1992). These hatch after 34 days at 30°C.

Key to *Diplodactylus* species

1 Nostril in contact with rostral	2
Nostril separated from rostral	**Western saddled ground gecko** (*pulcher*)

2 Enlarged apical lamellae followed by secondary series of enlarged lamellae	3
Enlarged apical lamellae followed by small granular lamellae for remainder of digit	**White-spotted ground gecko** (*alboguttatus*)

3 Dorsal pattern includes pale vertebral stripe or series of vertebral blotches	**Wheatbelt stone gecko** (*granariensis*)
Dorsal pattern lacks vertebrally aligned stripe or blotches	**Speckled stone gecko** (*polyophthalmus*)

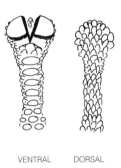

VENTRAL   DORSAL

Toe structure for *Diplodactylus* species

LIZARDS

White-spotted ground gecko
Diplodactylus alboguttatus Werner 1910

Genus Diplodactylus
Gray 1832

A large genus confined to Australia. Dorsal and ventral scales granular, smooth and homogeneous in size. Labials larger than adjacent scales. Digits not dilated at base, round, moderately expanded distally, corresponding to a pair of enlarged subdigital pads. Scales above distal expansion no larger than those above proximal parts of digit. All digits with retractile claw between apical pads. Preanal pores present or absent.

City Beach

HABITAT	Coastal and near-coastal dunes, limestone and inland sandplains supporting eucalypt or banksia woodlands and heath. Uncommon in the Perth area, with most recent records coming from Victoria Park, Bold Park and between Burns Beach and Yanchep. This region is the southern limit of its distribution.
DESCRIPTION	Nostril almost invariably contacts rostral. Enlarged apical lamellae followed by small granular lamellae for remainder of digit. Postanal tubercles 2–4. A single preanal pore located on each side in males. Ground colour pale yellowish to reddish-brown. A series of pale, dorsal blotches extend from top of head to tail. Sides with cream to white spots. Regrown tail peppered with fine black spots. Adult SVL 55 mm. Tail 80–110% of SVL.
GENERAL	Shelters in vertical shafts of abandoned spider burrows. Rarely found beneath surface debris (Wilson & Knowles, 1988).

Wheatbelt stone gecko
Diplodactylus granariensis Storr 1979 subsp. *granariensis*

Walyunga

HABITAT	Generally restricted to the Darling Range including the lower western foothills, although there are old records from the Swan Coastal Plain, Wanneroo and Subiaco. It seems to prefer hard ground beneath open vegetation. Locally, this gecko is less common than its close relative *D. polyophthalmus*.
DESCRIPTION	Similar to *D. polyophthalmus*, but has a dark-edged vertebral stripe or series of blotches and is usually greyish rather than reddish. Nostril almost invariably contacting rostral. Enlarged apical lamellae separated from a single series of moderately enlarged oval lamellae by two or three (rarely four or five) transverse rows of small granular scales. Postanal tubercles 3–9. Preanal pores absent. Ground colour brownish to greyish-brown. An unbroken (very rarely broken in the Perth region) pale vertebral stripe with straight or serrated edges. Belly white to grey. Adult SVL 50 mm. Tail 60–80% of SVL.
GENERAL	It is often found concealed beneath stones on soil, where it is able to excavate a short burrow. During winter it is not uncommon to find two or more individuals sharing the same site. Lays two soft-shelled eggs measuring roughly 16 mm long by 8 mm wide.

LIZARDS

Speckled stone gecko
Diplodactylus polyophthalmus Günther 1867

Above: Mount Dale   Below: Dianella

HABITAT   Generally found in rocky areas. Locally common in the Darling Range on laterite and less common on granite. Uncommon on the Swan Coastal Plain, where it is mostly restricted to the northern suburbs of Hillarys, Wanneroo and Eglinton, although it does occur on limestone in the Fremantle area and in eucalypt or banksia woodlands at Kings Park.

DESCRIPTION   Similar to *D. granariensis* but lacks dark-edged vertebral stripe or blotches and is usually reddish-brown. Nostril in contact with rostral. Moderately enlarged pair of apical lamellae, separated from a single row of round lamellae by 2–4 transverse rows of very small granular scales. Postanal tubercles 5–13. Preanal pores absent. Ground colour light to rich reddish brown on Darling Range and greyish-brown on coastal plain, with numerous pale spots. These may be so large as to reduce the ground colour to a fine reticulum. Belly white to brown. Adult SVL 45 mm. Tail 50–70% of SVL.

GENERAL   Usually found concealed beneath stones, bark and logs. Lays two soft-shelled eggs measuring roughly 15 mm long by 7 mm wide.

Western saddled ground gecko
Diplodactylus pulcher (Steindachner 1870)

Mount Dale

HABITAT This gecko is uncommon in the Perth area, where it is restricted to a few scattered localities such as Mount Dale in the Darling Range. In the wheatbelt north and east of this area, it is common beneath stones on granite soils.

DESCRIPTION Nostril separated from rostral by anterior nasal scale. Moderately enlarged pair of apical lamellae separated from a double row of round granular lamellae by 3–6 transverse rows of very small granular scales. Postanal tubercles 3–14. Preanal pores absent. Ground colour light brown, purple-brown to rich red-brown. A dark-edged pale vertebral stripe or series of blotches continuous with a pale head. Flanks with small pale spots. Belly white. Adult SVL 50 mm. Tail 40–60% of SVL.

GENERAL Believed to feed almost exclusively on termites. Lays two soft-shelled eggs measuring roughly 16 mm long by 8 mm wide. Eggs incubated in the laboratory hatched after 39 days at 30°C. Hatchlings measured 27–28 mm total length and weighed 0.57 g (Bush, 1992).

LIZARDS

Variegated dtella
Gehyra variegata (Duméril & Bibron 1836)

Genus *Gehyra*
Gray 1834

This genus comprises about 30 species, of which about half occur in Australia. Labial scales much larger than adjacent scales. Digits moderately long, depressed and greatly expanded distally, corresponding to a series of equal-sized, undivided or divided, narrow, transverse lamellae. Inner digit on each foot clawless. Claws on other digits free (not retractile) and arising from upper part of distal expansion. Preanal pores present.

Mount Dale

HABITAT — Common in the Darling Range on granite, but absent from the Swan Coastal Plain.

DESCRIPTION — Like *Phyllodactylus marmoratus*, but lacks a claw on inner digit of each foot. Scales above distal expansion of digits small (not noticeably larger than those before expansion), whereas in *Phyllodactylus marmoratus* they are large and plate-like (much larger than the scales before distal expansion). Dorsal and ventral scales granular, smooth and homogeneous. For further description see notes on genus. Preanal pores in males 9–18. Ground colour grey to purplish-brown, marked with blackish transverse bars and blotches, each followed by a smaller whitish spot. Side of head often with 1–3 longitudinal dark bars from eye to neck. Adult SVL 50 mm. Tail 90–130% of SVL.

GENERAL — Further east, this gecko is often found on standing and fallen trees, but locally it seems to be most common on granite. It may be excluded from trees in areas where this microhabitat is occupied by *P. marmoratus*, an ecological equivalent. Several individuals may be found beneath one rock. Lays a single brittle-shelled egg measuring roughly 10 mm long and 9 mm wide. Large numbers of failed eggs may be found beneath rocks and logs around the periphery of granite outcrops in the wheatbelt.

Binoe's prickly gecko
Heteronotia binoei (Gray 1845)

Genus *Heteronotia*
Wermuth 1965

A genus endemic to Australia, comprising three species, one introduced to this area. Labial scales much larger than adjacent scales. Dorsal scales heterogeneous: small and granular, interspersed with numerous trihedral tubercles, often arranged longitudinally. Toes long and bird-like with fixed claw, not expanded distally. Tail long, tapering and round in cross-section. Original tail scalation consists mainly of longitudinal rows of trihedral tubercles. Preanal pores present in males.

Paynes Find

HABITAT — One of the least habitat-specific geckos in Australia. This probably accounts for its success in colonising new areas like Perth. Until recently there was a population at Buckland Hill, Mosman Park, where it sheltered beneath limestone rocks and rubbish. It was probably inadvertently transported to this area in produce. A very common and widespread gecko found over much of Australia.

DESCRIPTION — See notes on genus for more details. Ground colour dark brown to reddish-brown with whitish, dark-edged irregular cross-bands, 1–3 scales wide, on body and tail. Interspaces twice as wide as bands. Adult SVL 50 mm. Tail 120–150% of SVL.

GENERAL — A ground-dwelling lizard that may be found beneath anything that will afford it cover. Appears to tolerate quite high temperatures and has even been observed basking. Both bisexual and all-female populations occur. Lays one or two brittle-shelled eggs measuring roughly 9 mm long by 8 mm wide.

LIZARDS

Reticulated velvet gecko
Oedura reticulata Bustard 1969

Genus *Oedura*
Gray 1842

A genus endemic to Australia, with about 13 species. All are tree or rock dwellers with depressed bodies. Digits depressed and moderately expanded distally, corresponding to a pair of enlarged apical lamellae and divided secondary lamellae. Scales on upper part of distal expansion no larger than scales on upper proximal parts. All digits with retractile claw between apical lamellae. Preanal pores present.

Walyunga

HABITAT In this area, restricted to the Darling Range. Uncommon.

DESCRIPTION Similar to *Phyllodactylus marmoratus*, but scales above distal expansion of digits no larger than those above proximal parts (whereas in *P. marmoratus* the scales above distal expansion are much larger than those above proximal parts). Dorsal scales small, granular and homogeneous; much smaller than ventral scales. Enlarged squarish apical lamellae separated from a few broadly oval lamellae by a divided series of 2–5 squarish lamellae. One postanal tubercle on each side. Preanal pores 2–5 in males. Dorsally greyish-brown with a broad pale vertebral zone or series of paravertebral blotches extending onto head and tail (unless regenerated). Dark stripe extending from nostril back through eye and along lower margin of pale vertebral zone. Remainder of body, legs and tail peppered with whitish spots. Belly pale grey. Adult SVL 65 mm. Tail 60–100% of SVL.

GENERAL An arboreal gecko found on smooth-barked eucalypts: locally wandoo (*Eucalyptus wandoo*), powderbark (*E. accedens*) and bullich (*E. megacarpa*). Rarely encountered on rough-barked trees. Outside this area it prefers salmon gum (*E. salmonophloia*) and gimlet (*E. salubris*). It has been observed feeding on sap oozing from trees (Dell, 1985).

Marbled gecko

Phyllodactylus marmoratus (Gray 1845) subsp. *marmoratus*

Genus *Phyllodactylus*
Gray 1828

A large genus represented in Australia by only two species. Digits moderately depressed with distal expansion corresponding to a pair of greatly enlarged apical lamellae. Two or three rows of small granular lamellae separate the apical lamellae from a single series of transversely oval lamellae. Scales on upper surface of distal expansion much larger than those on upper proximal parts of digit. All digits with retractile claw located between enlarged scales above apical lamellae. Preanal pores absent.

Shenton Park

HABITAT — Found throughout this area including Rottnest and Garden Islands. Still occurs in the long-developed older suburbs, where it forages at night on trees and walls. It does occur on coastal limestone, but in the Darling Range it is uncommon on granite, usually being found on standing and fallen trees. It is probably excluded from the granite because of the abundance of *Gehyra variegata*. In areas to the south outside the range of *Gehyra variegata*, it is very common on granite.

DESCRIPTION — Similar to *Gehyra variegata* but has claws on all digits. Dorsal and ventral scales granular, smooth and homogeneous. For further description see notes on genus. Ground colour grey, purplish-brown to (rarely) reddish-brown with narrow blackish-brown zig-zagging transverse bands or blotches, each followed by a contrasting pale blotch. Original tail, especially in juveniles, sometimes with distinct orange-red median blotches that may extend onto back. Belly pale grey. Adult SVL 65 mm. Tail 90–140% of SVL.

Duncraig

GENERAL — Lays two brittle-shelled eggs measuring roughly 10 mm long and 9 mm wide. These will incubate in the laboratory in dry sand (whereas leathery-shelled eggs incubated in a dry environment often fail because of dehydration). Hatchlings measure: SVL 13 mm, total length 28 mm, weight 0.08 g.

South-western spiny-tailed gecko *Strophurus spinigerus* (Gray 1842)

Genus *Strophurus* Fitzinger 1843

A genus comprising 10 species of tail-squirting gecko, until recently included in the genus *Diplodactylus*. Wells and Wellington (1983) resurrected the name *Strophurus*, but it is only recently that it has been widely accepted. Dorsal scales heterogeneous, with many large conical and spinose scales scattered amongst small flat granular scales. Labials much larger than adjacent scales. Digits not dilated at base, round, expanded distally, corresponding to a pair of enlarged subdigital pads. Scales above distal expansion no larger than those above proximal parts of digit. All digits with retractile claw between apical pads. Preanal pores present or absent.

Eastern subsp. *inornatus* (Storr 1988)   Mundaring

HABITAT Two subspecies found locally in a variety of habitats, from coastal dunes and heathlands, including Rottnest and Garden Islands (subsp. *spinigerus*), to the heavily timbered Darling Range (subsp. *inornatus*). Although *inornatus* is generally restricted to the Darling Range, it has recently been found at Ellenbrook and near Midland.

DESCRIPTION Dorsal scalation includes an intermittent or continuous paravertebral series of enlarged conical to spinose scales, which may or may not continue onto tail as large spinose tubercles. Nostril in contact with divided rostral. Enlarged pair of apical lamellae followed immediately by an undivided (rarely divided) series of moderately enlarged transversely elliptical lamellae. Postanal tubercles 2–5. Ground colour grey, finely peppered with black-tipped scales. In *spinigerus*, the black-tipped scales

Swan Coastal Plain subsp. *spinigerus* — Ellenbrook

are concentrated along the vertebral zone forming a broad vertebral stripe or series of rhomboidal blotches. This pattern is absent or poorly developed in *inornatus*. The tail is often densely peppered with black between the two rows of spines. Belly pale grey finely peppered with black, with or without a dark reticulum. Iris usually yellow locally in *spinigerus*, orange in *inornatus*. Adult SVL 75 mm. Tail 55–80% of SVL.

GENERAL
It is a common inhabitant of residential gardens and a terrestrial and arboreal forager that rests during the day in low to tall shrubs. It has also been observed resting on weldmesh gates. This is the only local gecko capable of squirting harmless, mildly odorous fluid from pores along the top of its tail. This is believed to deter predators. Lays two soft-shelled eggs, which measure roughly 16 mm long and 12 mm wide. Hatchlings SVL 20 mm.

Thick-tailed or barking gecko
Underwoodisaurus milii (Bory 1825)

Genus
Underwoodisaurus
Wermuth 1965

A genus endemic to Australia comprising two species. Labial scales much larger than adjacent scales. Preanal pores absent. Tail moderately long (not as long as body), broad proximally and tapering abruptly distally.

Neerabup

HABITAT	Very common in the Darling Range on laterite, granite outcrops and beneath rubbish, especially corrugated iron over laterite. Uncommon on coastal limestone.
DESCRIPTION	Dorsal scales heterogeneous: small granular scales interspersed with larger conical scales. Toes long, clawed and bird-like with no distal expansion. The tail is broad and flat nearest the body, tapering abruptly towards the tip. Ground colour reddish-brown to purplish-brown with white to yellowish spots, often arranged to form transverse bands. Original tail with white bands, but regrown tail usually monotonal purplish-brown. Adult SVL 95 mm. Tail 60–80% of SVL.
GENERAL	This gecko seems to be more cold-tolerant than other members of the family found locally. It may be the only species observed active on very cool nights. When disturbed it is quick to attain a defensive stance with the body raised off the ground, mouth agape and tail waving. If further provoked, it will leap forward and simultaneously emit an audible 'bark' in an attempt to defend itself. At this point it may dismember its tail or inflict a harmless bite. Lays two soft-shelled eggs.

Legless lizards
(Family Pygopodidae)

THIS family, also commonly called pygopods or pygopodids, is unique to Australia, with the exception of the genus *Lialis*, which also occurs in New Guinea. There are about 35 species known at present. More than half are found in Western Australia and many of these are endemic.

These lizards have evolved such that their front limbs are non-existent, and their rear limbs greatly reduced. All that remains of the rear limbs is a pair of hindlimb flaps located on each side at the base of the tail. These serve little purpose as limbs, although some pygopods have been observed using them to assist when climbing through grass tussocks. In the genus *Aprasia* the flaps are very small and probably no longer serve any function at all. They all lack eye lids, the eyes being protected by an immovable transparent spectacle, which is renewed with the shedding of each old skin.

The legless lizards are non-venomous, but are often killed because of their superficial resemblance to snakes. There are several external features that distinguish pygopods from snakes.

Most, especially the larger species, have a visible ear-opening on each side of the head behind the eyes, and a broad tongue that is often used to clean the lips and eyes. In contrast, snakes lack ears and have a long slender forked tongue.

The tail length in legless lizards ranges from slightly shorter than the body to four times body length, and the tail can easily break off at any point. A regrown tail can be recognised, as it will usually be of a different colour and pattern than the original. The ventral or belly scales of legless lizards are divided into two or more rows, and are included in the count of the midbody scales used in the following descriptions to assist identification. Venomous land snakes have a single row of wide ventral scales.

Many pygopod species vocalise with a high-pitched squeak when distressed. This characteristic is shared with the geckos, which are probably their closest relatives. All are egg layers, usually producing two eggs in spring and/or early summer. The eggs are deposited in soil beneath a rock or log. Pygopods vary greatly in appearance and habitat preference. The Worm Lizards (genus *Aprasia*) are small invertebrate feeders that forage in soil and leaf-litter and are seldom seen active on the surface. The larger species such as the Burton's Legless Lizard (*Lialis burtonis*) feed on small skinks and are semi-arboreal, climbing up and into tussock grass clumps and shrubs.

Key to Pygopodidae genera

| 1 | Scales on top of head large | 2 |
| | Scales on top of head small and fragmented | *Lialis* |

| 2 | Scales in 14 or fewer rows at midbody | *Aprasia* |
| | Scales in 15 or more rows at midbody | 3 |

| 3 | Scales on body smooth | 4 |
| | Scales on body keeled | *Pletholax* |

| 4 | Preanal pores present | *Pygopus* |
| | Preanal pores absent | 5 |

| 5 | Ventral scales larger than adjacent dorsals | *Delma* |
| | Ventral scales no larger than adjacent dorsals | *Aclys* |

Javelin legless lizard
Aclys concinna Kluge 1974 subsp. *concinna*

Genus *Aclys*
Kluge 1974

A unique genus endemic to Western Australia, represented by a single species (two subspecies). Restricted to the coast between Perth and Shark Bay, favouring dune and sandplain heath-lands. A moderately long, very slender genus with tail about four times body length. Snout pointed, head with enlarged scales and conspicuous ear-openings. Midbody scales in 20 rows, hindlimb flaps obvious, five scales along lower margin. Diurnal, very agile and semi-arboreal. Preanal pores absent.

G. Harold   Gairdner Range

HABITAT — Uncommon. Favours the white coastal dune and sandplain country north of the Swan River, where it occupies areas of low dense vegetation in banksia or eucalypt woodlands. Absent from the Darling Range.

DESCRIPTION — Body slender with smooth scales. Head scales enlarged. Snout long and pointed. Hindlimb flaps well developed. Midbody scales in 20 rows. Ground colour pale-grey with a broad dark-grey dorsal stripe extending from top of head to tail. A faint lateral line often extends along each side of anterior portion of body. Lips white. Belly white to pale grey. Adult SVL 10 cm. Tail up to four times SVL.

GENERAL — The first specimen collected (holotype) was found in a Sorrento backyard in 1962. Its name is derived from both its body shape and its habit of flinging itself into the air when startled. This behaviour, along with its speed and agility, make it very difficult to capture. Little is known about its habits, although it has been observed active in vegetation two metres above the ground.

Western granite worm lizard
Aprasia pulchella Gray 1839

Worm lizards
Genus *Aprasia*
Gray 1839

This genus comprises the smallest legless lizards; they are all less than 20 cm long. Scales smooth and shiny, in 12 or 14 rows at midbody. All have an overshot top jaw (more pronounced in certain species), which gives the snout a shovel shape when viewed from the side. This characteristic is an adaptation for burrowing or fossorial habits. The hindlimb flaps are minute compared with other legless lizard genera, with one or two scales along lower margin. All species lack external ear-openings, except one from Eastern Australia. Of the 11 or so species known, six are endemic to Western Australia. Preanal pores absent.

York

HABITAT — Common in woodland areas in the Darling Range, showing a preference for outcrops of granite and laterite. Occasional recordings from Swan Coastal Plain are probably specimens transported there accidentally in soil and firewood.

DESCRIPTION — Similar in many respects to *Aprasia repens*, but has 14 rather than 12 midbody scale rows and darker coloration. Snout and face dark greyish brown, back of neck and anterior portion of body rusty brown, merging to grey base colour along body and tail. Each dorsal and lateral scale is marked in its centre with a tiny black dash, which together form longitudinal series of stripes from neck, becoming more distinct on tail. Lips white, ventral surface white to pale-brown on belly, white below tail. Adult SVL 12 cm. Tail slightly shorter than SVL and terminating in blunt tip.

GENERAL — A fossorial lizard, endemic to Western Australia. It favours rocky areas, where it forages among leaf-litter and other ground debris. Feeds on small invertebrates, presumably termites and ant eggs. Shelters in moist soil beneath rocks and in rotten logs. When handled it often emits a faint, high-pitched squeak.

South-western sandplain worm lizard *Aprasia repens* (Fry 1914)

Northam

HABITAT
: Common on the Swan Coastal Plain in most habitat types. Also found on Rottnest Island. Less common in the Darling Range.

DESCRIPTION
: A small slender species similar in most respects to *Aprasia pulchella* but having a more protrusive upper jaw and 12 rather than 14 midbody scale rows. Light brown to light grey above. Scales are usually dark-centred, forming a series of longitudinal spots or lines along body, usually more pronounced on the tail. Lower chin and throat yellow, ventral surface whitish often yellow towards tail. Adult SVL 110 mm. Tail slightly shorter than SVL.

GENERAL
: Forages in loose soil and overlying leaf-litter for small invertebrates. A proficient burrower, which shelters in soil beneath rocks and logs. When handled may emit a high-pitched squeak. Endemic to Western Australia.

Fraser's legless lizard

Delma fraseri Gray 1831 subsp. *fraseri*

Genus *Delma*
Gray 1831

A large genus of legless lizards found throughout most of Australia, from coastal heaths and forests to arid inland habitats. Diurnal and nocturnal. Small to moderately large, with a relatively long tail, up to four times SVL. Scales smooth and shiny, midbody scale rows 12–20. Head scales enlarged. Ear-openings obvious. Hindlimb flap obvious, 2–6 scales along lower margin. About 20 species are currently known, half of them occurring in Western Australia. Preanal pores are absent.

Wanneroo

HABITAT	Common on the Swan Coastal Plain, more so north of the river than south. Found in a range of habitat types including coastal heath, scrubland, jarrah and banksia woodlands. Moderately common in the Darling Range, generally in woodlands.
DESCRIPTION	A large legless lizard with obvious ear apertures and large hindlimb flaps. Body scales are smooth and may have a slight shine. Midbody scales in 15–17 rows, usually 16. Ground colour reddish brown to grey, merging to olive-brown towards tail. Head and nape dark-brown to black above, very distinct in juveniles, fading in adults. There are three black bars on each side of the head between ear and tip of snout, which extend onto throat. Ventral scales have dark margins. Adult SVL 12 cm.
GENERAL	Forages in low vegetation. Shelters beneath rocks, logs and fallen bark, also commonly found beneath discarded fibro and corrugated iron sheets. When disturbed it thrashes frantically to avoid capture. It can be mistaken for a baby Dugite because of its dark head. Diurnal, very agile and semi-arboreal. Bush (1984b) observed large numbers aggregating together during winter. Egg-laying recorded in December. Egg size 22–23 mm x 7–8 mm, weight 1.04–1.14 g. Hatched after 74–77 days at 28°C. Neonate SVL 43–45 mm, weight 0.72–0.96 g (Bush, 1983a).

Gray's legless lizard
Delma grayii Smith 1849

Neerabup

HABITAT	Moderately common in coastal dunes covered by low heath, and on coastal sandplain, where it prefers areas dominated by banksia and blackboy. Uncommon in the Darling Range.
DESCRIPTION	A moderately large and slender legless lizard with large symmetrical head scales. Midbody scales in 15–18 rows. Top of head and upper surface of body pale-grey to olive-brown, often with regularly spaced dark spots. Flanks usually tinged with coppery brown and often more densely spotted than back. Sides of head (behind eyes) and anterior portion of body marked with white or yellow vertical bars. Distinguished from *D. fraseri* by having yellow lips, chin and belly. Pale-grey below tail. SVL up to 11 cm. Tail up to four times SVL.
GENERAL	Usually diurnal, but has been observed active at night. Shelters in and beneath dead vegetation. Retreats underground to become inactive during winter. Endemic to Western Australia.

LIZARDS

Burton's legless lizard
Lialis burtonis Gray 1835

Langford

Genus *Lialis*
Gray 1835

This genus contains two species, one occurring extensively in Australia, southern and eastern New Guinea, Torres Strait and Aru islands. The other is restricted to New Guinea and adjacent islands. They are large and moderately stout legless lizards with distinctly long and narrow wedge-shaped snouts. External ear-openings present. Hindlimb flap small, 1–3 scales along lower margin. Head scales small and fragmented, large plate-like scales absent. Eye small with elliptical pupil. Preanal pores present. The only pygopodids that feed exclusively on other reptiles.

HABITAT — Very common throughout the Perth region, including Rottnest and Garden Islands. Found in most habitat types but especially in areas of low vegetation such as scrubland and heath. Also known from forested and rocky areas of Darling Range.

DESCRIPTION — A large legless lizard with long pointed snout. Ear aperture and hindlimb flaps small. Pupil vertically elliptical. There are 18–22 midbody scale rows. Ground colour highly variable, from uniform pale-grey, cream, tan or rusty brown to orange, with or without a series of black longitudinal dashes extending along body and tail. These may be fused as lines. Often a white stripe extends through lips, becoming obscure along lower flanks. Belly colour may be the same as body or much darker with tiny pale specks. Adult SVL up to 30 cm. Tail 1–1.5 times as long as SVL.
(Continued on next page)

Burton's legless lizard
(Continued)

Mandurah

GENERAL — Diurnal during the cooler autumn and spring months, becoming nocturnal on warm summer nights. Preys on small lizards, which it catches in or under low vegetation using a stealth-strike method. The prey is grasped firmly around the neck and swallowed head first. Feeding is often followed by cleaning of snout and eyes with the tongue. Cannibalistic at times, feeding on other legless lizards and occasionally small snakes. Shelters beneath rocks, logs, fallen bark and rubbish. As with all other legless lizards, this species is harmless. Egg-laying recorded in January. Egg size 22–23 mm long and 9–11 mm wide, weight 1.8–2 g. Hatched after 66–68 days at 28°C. Neonate SVL 66–68 mm, weight 1.5–1.7 g (Maryan, 1987).

Keeled legless lizard

Pletholax gracilis Cope 1864 subsp. *gracilis*

Genus *Pletholax*
Cope 1864

A monotypic genus restricted to the mid and lower west coast of Western Australia. A medium-sized, very slender, long-tailed legless lizard. External ear-openings minute. Hindlimb flap very small, 1–2 scales along lower margin. Body scales including ventrals bear multiple keels, a unique characteristic in the pygopodids. Midbody scales in 16 rows. The ventral scales are not noticeably wider than adjacent dorsal scales. Tail less fragile than in other genera. Preanal pores are absent.

City Beach

HABITAT — Uncommon in the Perth region. Inhabits coastal dunes and sandplain on white sand supporting heath and banksia woodland. Recorded from Red Hill in the Darling Range. Endemic to Western Australia.

DESCRIPTION — Body very slender. All scales including ventrals having two distinct keels. Head scales large and symmetrical, snout long. Midbody scales in 16 rows. Hindlimb flaps and ear apertures small. Dorsally grey, top and sides of head dark-grey to almost black, merging to rusty or greyish brown along sides of body and tail. Lips, throat and forebelly yellow, whitish beneath tail. Adult SVL 80 mm. Tail up to four times SVL.

GENERAL — This species is unique among legless lizards in having keeled belly scales. Little is known about its habits, although captive specimens have been observed burrowing into sand. This is unusual, as fossorial or burrowing habits are not normally associated with keeled scales. Further observations may show semi-arboreal habits, burrowing only for shelter during winter months. Individuals released by Brian Bush after removal from pit-traps, moved off the ground into low bushes. Diurnal, usually associated with low dense vegetation. (For further notes on its biology, see Shea & Peterson, 1993.)

Common scaly foot
Pygopus lepidopodus (Lacépède 1804)

Genus *Pygopus*
Merrem 1820

Contains two species that occur throughout most of mainland Australia. They are large stout legless lizards with a blunt rounded snout. External ear-openings conspicuous. Hindlimb flaps large, 3–10 scales along lower margin. Midbody scales in 21 or more rows. Dorsal scales (rarely) smooth to strongly keeled. Large cloacal spur in males. Preanal pores present.

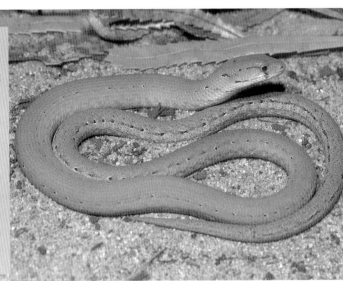

Neerabup

HABITAT	Moderately common on Swan Coastal Plain, preferring sandy soils supporting shrublands and other low dense vegetation. We are aware of only two specimens recorded from the Darling Range, both found near Karragullen.
DESCRIPTION	A relatively large, heavily built legless lizard with short, rounded snout and strongly keeled dorsal scales. The hind limb flaps are large and ear-openings are conspicuous. There are 21–25 midbody scale rows and 12–18 preanal pores. Ground colour grey to reddish brown, the degree of patterning is variable. It can be absent or in the form of longitudinal dashes, which may be faint or strikingly distinct and continuous along body. Distinct pattern, when present, consists of three dorsal stripes composed of black bars interspaced longitudinally by reddish-brown and edged with white. When these stripes are present, the ground colour is invariably grey. Belly pale-grey, brown or pink, with or without darker variegations. Highly patterned specimens occur more frequently outside the Perth region. SVL up to 230 mm. Tail up to twice SVL.

Note ear-opening   Lort River

GENERAL   Active during the day in spring and autumn, becoming nocturnal in summer. Feeds on spiders and insects. When threatened it displays snake-like behaviour, raising the head and neck and waving protruded tongue. When handled it will squeak and twist violently between the fingers to escape. This often results in the tail breaking. The Scaly Foot has the largest hindlimb flaps of all the legless lizards. Because of this feature it is generally considered more primitive (goes further back along the evolutionary path) than its relatives with smaller hindlimb flaps. Lays one (Daly, 1992) or two (Bush, 1992) eggs in December. These measure 40mm long and 17mm wide, and hatch after 70–110 days, depending on incubation temperature. Communal egg-laying has been reported in this species (Bush, 1981).

Dragon lizards
(Family Agamidae)

THIS family is represented in Western Australia by more than 45 species, of which three occur locally. Another species of dragon lizard that may occasionally be found in Perth, although it does not occur here naturally, is the Thorny Devil (*Moloch horridus*). People sometimes pick up these strange-looking lizards while on holidays to the north or east of Perth, and bring them back as pets. These lizards are specialised feeders, requiring large numbers of ants, and generally do not thrive in captivity.

Dragon lizards feed on invertebrates, smaller reptiles and some vegetable matter. All are egg-layers and almost exclusively diurnal (daytime active).

The best-known member of this family is the Frilled Lizard (*Chlamydosaurus kingii*), which is restricted to northern Australia (in Western Australia, it is restricted to the Kimberley).

LIZARDS

Key to Agamidae genera and species

1 Conspicuous transverse row of spines
across back of head — **Western bearded dragon** (*Pogona minor*)

No transverse row of spines across
back of neck — 2

2 Small (total length less than 12 cm), with
series of spines along both sides of base
of tail. Generally found on sand — **Western heath dragon** (*Tympanocryptis adelaidensis*)

Large (total length up to 30 cm), without
spines on base of tail. Generally found on
granite outcrops and ledges — **Ornate crevice dragon** (*Ctenophorus ornatus*)

Moloch horridus,
Thorny devil,
Kambalda

P. Orange

Ornate Crevice Dragon
Ctenophorus ornatus (Gray 1845)

Comb-bearing dragons
Genus *Ctenophorus*
Fitzinger 1843

A diverse group of more than 22 species, only one of which is found locally. The common name for this group probably refers to the comb-like fringes of the eyelids. This genus attains its greatest diversity in arid and semi-arid areas. Comb-bearing dragons range in size from very small (SVL 50 mm) to moderately large (SVL 120 mm). All but one species have numerous femoral and preanal pores. The different species vary from having marginally compressed bodies to strongly depressed bodies, their scales are homogeneous to heterogeneous and they vary from being terrestrial to rock-dwelling.

Female — Red Hill

HABITAT — Although occasionally found away from rock, this species is most often observed on granite outcrops in the Darling Range. It is absent from the Swan Coastal Plain.

DESCRIPTION — Head and base of tail noticeably depressed. Gular scales smooth. Nuchal crest absent or very low. Dorsal scales homogeneous. Femoral and preanal pores 22–35 each side. Dorsum in males black to dark-brown, in females grey to pale-brown. A series of pale blotches along vertebral zone, remainder of dorsum with transverse pale spots increasing on lower flanks. Tail and limbs with pale bands. Black chest-patch in males. Adult SVL up to 95 mm.

LIZARDS

Male   Red Hill

GENERAL

The depressed body of this lizard allows it to seek shelter in narrow horizontal crevices and beneath exfoliated (flat layers of) granite. Anyone exploring a large granite outcrop like Boulder or Sullivan Rock during the warmer months will see these dragons on the run. They are often observed head-bobbing and arm-waving, which is a behaviour common in many agamid lizards. Four eggs have been recorded in a clutch.

Western bearded dragon
Pogona minor (Sternfeld 1919) subsp. *minor*

Bearded dragons
Genus *Pogona* Storr 1982

These dragons are recognised by most people, with their bearded appearance and spiky bodies. The bearded dragon is often mistakenly referred to as a Frilled Lizard, which is restricted to northern Australia.

These dragon lizards attain up to 60 cm in total length, have large triangular-shaped heads and numerous tubercles and spines over the body. These are particularly concentrated at the back of the head and along the flanks. They display varying degrees of arboreal behaviour. There is only one species represented locally.

Kings Park

HABITAT	Found in a variety of habitats, from the coastal dunes to the heavily timbered Darling Range.
DESCRIPTION	Gular scales keeled, nuchal crest absent, dorsal scales heterogeneous. Transverse row of spines across back of head, largest below and behind ear. A single series of enlarged spines along each flank, separating the upper and lower parts of the body. Dorsum grey to yellowish-brown. A series of large pale blotches on either side of the vertebral line. Often a dark bar between the eye and ear followed by a dark blotch behind the ear. Adult SVL up to 150 mm.
GENERAL	A terrestrial dragon that is often observed basking on fallen timber and rocks. This lizard changes colour depending on mood and body temperature. At rest or when disturbed it is dark grey. It becomes yellowish-brown as its body temperature rises, usually after basking. Lays two clutches of 5–11 eggs between October and February. At this time of the year large numbers of females bask on roads, presumably to speed up egg development. As a result, many are killed by vehicles. Egg size 20–25 mm long and 12–14 mm wide, weight 1.85–2.86 g. Hatched after 72–82 days at 25°C and 45–54 days at 30°C. Neonate SVL 32–37 mm, weight 1.7–2.9 g (Bush, 1992).

Western heath dragon
Tympanocryptis adelaidensis (Gray 1841) subsp. *adelaidensis*

Earless dragons
Genus *Tympanocryptis*
Peters 1863

Many of the species assigned to this genus lack obvious ear-openings. They are small short-tailed terrestrial dragon lizards, measuring up to 75 mm SVL. They have stout, marginally depressed bodies and heterogeneous body scales. These include large spinose scales interspersed among numerous small and very small scales. All have cryptic habits. Their behaviour, colour and body shape all add to their ability to be hidden. Only one species occurs naturally in the Perth area.

A separate genus, *Rankinia*, has been proposed by Wells and Wellington (1983) and used by Greer (1989) for several species, including *T. adelaidensis*, currently assigned to this genus.

Female   Melaleuca Park

HABITAT Common in low coastal vegetation on beaches and dunes, including heathlands and banksia woodlands on the Swan Coastal Plain. Absent from the Darling Range.

DESCRIPTION Gular scales smooth to weakly keeled, nuchal crest absent, dorsal scales heterogeneous. Paravertebral row of enlarged keeled scales, sometimes forming spines on neck. Flanks often with enlarged keeled scales or spines, forming a continuous series with those on tail. Dorsum grey to brown. Wide unmarked vertebral zone bordered by dark blotches that are somewhat triangular in shape and widest along mid-line. A second laterodorsal series of blotches is less defined than those dorsally. Head and limbs with darker markings. Black markings on throat and chest more distinct in males than females. Adult SVL 55 mm.
(Continued on next page)

Western heath dragon
(Continued)

Male   Ellenbrook

GENERAL — A small agile dragon that moves in short rapid bursts when pursued. When stationary its cryptic colouring allows it to blend well with the litter-strewn ground. Four eggs have been recorded in a clutch. Smallest SVL (23 mm) and weight (0.45 g) was recorded in late February. Bamford (1992) presents data on growth and sexual size dimorphism.

Goannas or monitor lizards
(Family Varanidae)

THE goanna or monitor family is monogeneric, with all species belonging to the genus *Varanus* Merrem 1820. The family contains the world's largest lizards. Although the species vary in size, from 20 cm in the Short-tailed Monitor (*Varanus brevicauda*) to 2.5 m in the Perentie (*Varanus giganteus*), they have many similar characteristics that together distinguish them from other lizard families.

The snout and neck is long and the head somewhat flat. Their limbs are well developed, with long sharp claws used for climbing, digging or defence, depending on the species. The head and body scales are small and the skin is rough and loose-fitting. The tail may have flat to strongly keeled scales and may be slightly shorter to considerably longer than the body. It is non-fragile and provides assistance when climbing or swimming. Some larger species such as the Gould's Monitor (*V. gouldii*) will use the tail as a weapon, or for balance while standing erect on the back legs. Monitors are the only Australian lizards possessing forked tongues, which aid in the sense of smell. As the tongue is flicked in and out, it gathers tiny particles from the air and surrounding objects. These are taken inside the mouth to the Jacobson's organ located in the roof of the mouth, where they are analysed.

Monitors are diurnal, moving about during the day in search of food. Prey items include any animal of suitable size to swallow, some species also feed on bird or reptile eggs and carrion. The Gould's Monitor (*V. gouldii*) may be observed on outback roads feeding on the carcasses of kangaroos and sheep. It can be attracted by the air-borne scent of carrion from several kilometres up-wind. Small species and hatchlings feed mainly on insects and lizards.

All monitors have sharp teeth for catching and holding prey. These are also used for defence. A large specimen with sharp teeth and claws should be shown respect and only handled by an experienced person. Even so, an observer can stand within a few metres and quietly watch without any fear of alarming the animal. Monitors will not attack when startled, they will run for cover. Some species such as the Southern Heath Monitor (*V. rosenbergi*) are semi-arboreal, racing up a tree to avoid capture, others will readily take to water. They occupy a vast range of habitat types throughout Australia. This is not surprising, as the very large species have few natural enemies. Monitors are egg-layers. The young are more distinctly patterned than the adults.

Key to *Varanus* species

1 Tail tip distinctly yellow or white	**Gould's monitor** (*gouldii*)
Tail tip not distinctly coloured	2

2 Back pattern includes up to 16 black bands	**Southern heath monitor** (*rosenbergi*)
Back pattern without black bands and, at most, consists of pale-edged black spots arranged longitudinally	**Black-tailed monitor** (*tristis*)

LIZARDS

Gould's monitor
Varanus gouldii (Gray 1838)

G. Harold   Mount Dale

HABITAT — Moderately common on Swan Coastal Plain in most habitat types, preferring areas of sandy soil supporting open woodlands. Less common on laterite soils supporting jarrah woodlands of Darling Range.

DESCRIPTION — A large monitor with laterally compressed tail. Ground colour brown, dark-grey to almost black. A pale-edged black bar extends through eye, back towards neck. Neck patterned with pale cream or yellow ocelli and spots, which become aligned into pale transverse bands across back. Tail marked with pale-yellow or cream bands, interspaced by dark brown. End of tail distinctly yellow or white. Limbs dark with large pale spots. Belly pale-yellow or dirty white with irregular grey spots. Grey chevron on throat. Total length of 150 cm.

GENERAL — A ground-dwelling monitor, which digs a burrow for shelter. This may be beneath a large slab of rock or under rubbish such as corrugated iron. Forages during the day for reptiles, insects, mice and carrion. When approached slowly, will often remain unconcerned by human presence. A large individual may hold its ground using defensive bluff tactics such as hissing and raising the forebody. If startled it will run for cover with amazing speed, hence the name 'race horse goanna' for the larger monitors. Eggs deposited in a burrow and left to incubate.

Southern heath monitor
Varanus rosenbergi Mertens 1957

Adult   Jandakot

HABITAT   Uncommon throughout this region. Occasionally recorded on Swan Coastal Plain north and south of the Swan River in sandy localities supporting woodlands and shrublands. Sometimes encountered in the Darling Range. Perth represents the northern limit of its distribution; it is more common to the south-east.

DESCRIPTION   A moderately large monitor with tail laterally compressed. Ground colour dark-grey, peppered with pale-grey to pale-yellow. Top of head almost black. A black line extends through each eye, back towards neck. There are several dark bands on back of neck, which curve forward onto sides. Back patterned with dark transverse bands, with interspaces much wider than bands. Bands continue to tip of tail, unlike *V. gouldii*, which has unbanded yellow tip of tail. Limbs are black, spotted with cream to pale yellow. Ventral surface cream to pale-yellow with a darker net-like pattern and sometimes with faint transverse bands. As with all monitor lizards, juveniles are more brightly coloured than adults. Total adult length of 150 cm.

LIZARDS

Juvenile   Fitzgerald River

GENERAL   Spends most of its time on the ground, but if chased it may rapidly climb
           the nearest tree with the aid of its sharp claws and powerful limbs. Digs a
           burrow for shelter or may utilise burrows of other animals. Feeds on
           frogs, reptiles, birds, insects and carrion. King and Green (1979) found
           that on Kangaroo Island the eggs of this species may be laid in termite
           mounds during the summer—incubating over winter and hatching in the
           early spring. It has yet to be ascertained if this is the case locally.

Black-tailed monitor
Varanus tristis (Schlegel 1839) subsp. *tristis*

Adult — Darlington

HABITAT
: Moderately common from jarrah forest and granite outcrops in the Darling Range. Uncommon on Swan Coastal Plain, where it is occasionally found in woodland areas. Perth represents the south-western limit of its range; it is more common to the north and east of Perth.

DESCRIPTION
: A moderately large slender monitor with depressed head and long, strongly keeled tail. Coloration varies with age and locality; the following description applies to the southern form found locally. Head, neck and tail black. Legs black with grey or brownish spots. Body light-grey patterned with ocelli or spots, often with indistinct transverse zones of orange-brown across body. This pattern darkens with age and an old individual may become completely black. The head and neck in juveniles is grey to brown, with indistinct bands of transversely arranged ocelli extending from the neck to the base of tail, these are interspaced by dark brown. The tail has a series of light brown bands interspaced by dark brown bands, becoming indistinct towards tip. Total adult length to 80 cm.

Juvenile   Lesmurdie

GENERAL

A very alert and agile climbing monitor, which is able to ascend trees and scale steep rocks with ease, assisted by its long claws. It often takes up residence in the ceiling space in houses and may be seen basking on the roof or on erected structures nearby. Feeds on birds, eggs, frogs, lizards and insects. The dark colour of the southern population may be an adaptation that increases heat absorption in the cooler climate.

Skinks

(Family Scincidae)

THIS is a large worldwide family, with members occupying virtually all terrestrial habitats. Skinks are well represented in Australia and are our largest reptile family. The family contains more than 300 species in 25 or so genera, half of which are represented in Western Australia. Because of continual changes in taxonomy (the science of classification) and exhaustive research efforts through field work, new species are recognised regularly, particularly in the genera *Ctenotus* and *Lerista*.

The family encompasses an enormous variety of colours, sizes, shapes and habits. The majority of skinks found around Perth have smooth shiny body scales, but exceptions are the Bobtail (*Tiliqua rugosa*), which has large rugose scales, and some *Egernia* species, which have keeled scales. Many have a fragile tail that is easily dismembered during a struggle. This allows the lizard to escape, while the assailant is left with a wriggling tail. A new tail is soon regrown.

The limbs of the more heavily-built skinks are usually well developed, but the slender ones have greatly reduced limbs. The reduction in limb size is an evolutionary adaptation that has occurred in many burrowing forms. An example is the genus *Lerista*, where the limbs are completely absent in some species. Some skinks from sub-humid and arid areas have movable scaly eyelids, which often bear a transparent disc. These protect the eyes while the skink forages through soil. Some small genera such as *Menetia* and *Morethia* have a noticeably larger eye disc between fused eyelids, forming a transparent spectacle. It is thought that these fused eyelids help to reduce moisture-loss through the eye.

Small skinks feed mainly on insects, whereas some larger species have quite varied diets, consuming flowers, fruits and occasionally carrion. Most skinks lay soft-shelled eggs, though a small number of species bear live young. The live-bearing species usually inhabit the cooler region of southern Australia, where the incubation of eggs would be inhibited by lower temperatures. Some live-bearers provide placental nourishment from mother to young.

Key to Scincidae genera

1	Moderately large to very large skinks (SVL up to 310 mm). Parietals completely separated	2
	Small to moderately large skinks (SVL up to 125 mm). Parietals in contact behind interparietal	4
2	Head very distinct from neck, occipital scales present	*Tiliqua*
	Head not or hardly distinct from neck, occipital scales absent	3
3	Front edge of ear-opening lined with lobules. Fourth hindlimb digit longer than third	*Egernia* (see p. 123)
	Front edge of ear-opening without lobules. Fourth hindlimb digit no longer than third	*Cyclodomorphus*
4	Five forelimb digits	5
	Four or fewer forelimb digits	10
5	Lower eyelid immovable	6
	Lower eyelid movable	7
6	Supranasals present	*Morethia*
	Supranasals absent	*Cryptoblepharus*
7	Lower eyelid with transparent disc	8
	Lower eyelid lacks transparent disc	9
8	Ear-opening prominent	*Bassiana*
	Ear-opening absent	*Hemiergis*

9 Ear lobules conspicuous. Upper tail scales smooth. Body pattern consists of bold longitudinal stripes ***Ctenotus*** (see p. 113)

Ear lobules absent. Upper tail scales usually keeled. Body pattern consists of bold cross-bands ***Eremiascincus***

10 Four forelimb digits, five hindlimb digits ***Menetia***

Four or fewer forelimb digits, four or fewer hindlimb digits ***Lerista*** (see p. 130)

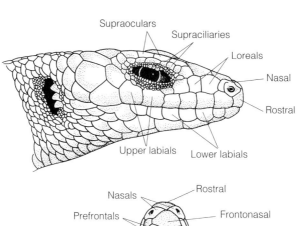

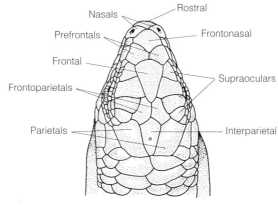

Head scales on skink lizards

South-western cool skink
Bassiana trilineata (Gray 1838)

Genus *Bassiana*
Hutchison, Donnellan, Baverstock, Krieg, Simms & Burgin 1990

Contains three species from temperate southern Australia, including Tasmania. These species were removed from the genus *Leiolopisma* in 1990. All are small to medium-sized terrestrial skinks with well-developed limbs. Each limb has five digits. Characterised by prominent ear-openings. Frontoparietals fused to form a single scale, interparietal very small and supranasals absent. Breeding males in most populations have red throat. Lower eyelid movable, with transparent disc. Scales with weak gloss.

South Perth

HABITAT	In this region generally associated with lakes, swamps, rivers, catchment areas and other damp localities, both on the Swan Coastal Plain and in the Darling Range. Occurs on Rottnest and Garden Island.
DESCRIPTION	A moderately robust, smooth-scaled skink with well-developed limbs, each with five digits. Eyelid movable, with transparent disc. Ground colour dark brown to dark grey. A black vertebral line and broad black lateral stripe from the neck to base of tail. The black lateral stripe is bordered above and below by white to pale-grey stripes. Lower areas on sides of body are grey, with each scale edged and/or flecked with black. Head sometimes tinged with coppery brown. Belly white to silver-grey. Adults of both sexes may have reddish chin and throat, the colour intensifying during the breeding season. Adult SVL 70 mm.

Ellenbrook

GENERAL   Sun-loving. Often seen basking on logs in the morning or actively foraging in grass and leaf-litter, hunting for soft-bodied invertebrates. Seeks shelter in dense vegetation and beneath leaf-litter and logs adjacent to wetlands. It lays eggs, producing clutches of up to seven.

Snake-eyed, fence or sun skink
Cryptoblepharus plagiocephalus (Cocteau 1836)

Genus *Cryptoblepharus*
Wiegmann 1834

A large genus, that is well-represented in other parts of the world. Six species currently recognised from Australia, with a single species being found locally. They are small, strongly depressed skinks with large eyes bearing immovable eyelids. The eyelids are fused to form a transparent spectacle, which is surrounded by small granular scales. Parietals and interparietal are fused into a single large scale.

Melaleuca Park

HABITAT	Abundant throughout the Perth region in most types of habitat. Favouring woodlands, rocky outcrops and man-made structures. Occurs on Garden Island
DESCRIPTION	A small skink with smooth glossy scales and five fingers and toes. The lower eyelid is immovable, forming a spectacle over the eye. The name Snake-eyed Skink is derived from this. Ground colour dark olive-grey with dark-brown or black spots. Usually some indication of a narrow pale dorsolateral stripe that extends from above the eye to the tail. Upper lateral area darker, with scattered spots that tend to coalesce into an upper lateral stripe. Belly whitish or silver-grey. Adult SVL 47 mm.
GENERAL	One of the commonest lizards in the Perth region. Prefers open vertical surfaces of trees and rocks. It has adapted successfully to living in developed areas, seeming comfortable on walls, fences and telegraph poles. Being very swift and agile, it is an efficient hunter of small insects, even leaping into the air to catch flies. It lays eggs, producing clutches of two or three.

Key to *Ctenotus* species

1	Back pattern absent or at most includes a black vertebral stripe	2
	Back pattern present and includes a series of 4–6 narrow longitudinal stripes	4

2	Hindlimbs boldly marked with black	3
	Hindlimbs finely peppered with black	**Darling Range heath ctenotus** (*delli*)

3	Lateral pattern includes dorsolateral series of bold white spots	**Jewelled ctenotus** (*gemmula*)
	Lateral pattern lacks white spots but includes unbroken white dorsolateral stripe	**Red-legged ctenotus** (*labillardieri*)

4	Upper lateral zone includes one or more series of white spots or blotches. Prefrontals usually in contact	5
	Upper lateral zone without white spots or blotches but consists of unbroken stripes. Prefrontals separated	**South-western odd-striped ctenotus** (*impar*)

5	Back with four white stripes, unbroken white mid-lateral stripe below a single series of white spots or blotches	**West coast ctenotus** (*fallens*)
	Back with six white stripes, lateral pattern without unbroken stripes but consists wholly of white spots, blotches and transverse bars	**Western limestone ctenotus** (*lesueurii*)

Darling Range heath ctenotus
Ctenotus delli Storr 1974

Comb-eared skinks
Genus *Ctenotus*
Storr 1964

Australia's largest reptile genus, containing more than 80 species and many subspecies. Further field work will no doubt reveal additional taxa. The greatest diversity of these skinks is found in arid to semi-arid regions, amongst hummock grasses.

These skinks are small to moderately large, slender, with long tapering tails. There are six species in this region. *(Continued opposite)*

Karragullen

HABITAT	Associated with jarrah and marri woodlands that have a shrub-dominated understorey on laterite, sand or clay soils in the Darling Range. Occasionally found on granite outcrops. Absent from the Swan Coastal Plain.
DESCRIPTION	Ground colour dark olive to dark coppery brown, unmarked except for a narrow black laterodorsal stripe and white dorsolateral stripe, which is broken to form a series of white dashes. These extend from above eye to base of tail. Upper lateral zone black, enclosing one or more series of scattered white dots. White midlateral stripe is represented by a series of short dashes. Limbs brown and finely peppered with black. Belly white to yellow. Adult SVL 63 mm.
GENERAL	Shelters beneath rocks, dense vegetation, dead blackboys and in abandoned stick-ant nests and shallow burrows.

LIZARDS

West coast ctenotus
Ctenotus fallens Storr 1974

Red Hill

The limbs are well developed, each bearing five digits. The eyes are relatively large with a movable lower eyelid. They have obvious ear-openings with anterior ear lobes. The body pattern is simple to complex, usually consisting of stripes or stripes in combination with series of spots.

All species are alert, fast-moving, terrestrial and diurnal. They forage mainly amongst low vegetation, rocks and other partial cover, feeding on a wide variety of invertebrates. They are egg-layers, producing small clutches (see key p. 113).

HABITAT   Moderately common throughout the Perth region. On the Swan Coastal Plain, occurs in most habitats supporting a shrub-dominated understorey, but favours low coastal vegetation on sandy soils. In the Darling Range prefers areas of granite and is found less commonly on laterite. Occurs on Rottnest Island.

DESCRIPTION   Similar to *C. lesueurii*, but its back pattern usually includes four narrow white stripes rather than six. Ground colour dark to pale greyish-brown or yellowish-brown in adults, and blackish in juveniles. Vertebral stripe blackish-brown to black with a white edge, which is in turn sometimes edged with black, extends from nape to base of tail. White dorsolateral stripe from eye tends to become diffused with brown or yellow on tail. Upper lateral zone dark brown enclosing a series of white blotches, spots or dashes. Limbs brown and striped with darker brown or black. Belly silver-white to yellow. Adult SVL 95 mm.

GENERAL   Agile and diurnal in the summer. During winter it is often found concealed in a shallow burrow beneath rocks or logs. Also found in disturbed areas with scattered rubbish and introduced grasses.

Jewelled ctenotus
Ctenotus gemmula Storr 1974

Ellenbrook

HABITAT — Scarce on the Swan Coastal Plain, where it inhabits pale soils supporting heathlands, usually in association with banksia. This species seems only to occur in isolated populations in south-west Western Australia, reaching its northern limit in the Perth region.

DESCRIPTION — Ground colour pale silvery brown to yellowish brown. Broad black upper lateral zone extending marginally onto back. Usually with an upper series of prominent white spots and a wavy white mid to lower-lateral stripe or series of dashes. Limbs yellowish brown, boldly marked with black and white. Belly white. Adult SVL 58 mm.

GENERAL — Seeks shelter beneath leaf-litter, in abandoned stick-ant nests and burrows at the base of banksia trees and shrubs.

South-western odd-striped ctenotus *Ctenotus impar* Storr 1969

Mandurah

HABITAT

On the Swan Coastal Plain, it seems to be most common south of the Swan River. Inhabits sandplain supporting heath in association with banksia woodlands. In the Darling Range it prefers areas of heath with scattered eucalypts on sandy soils.

DESCRIPTION

A distinctly striped ctenotus with 11 (occasionally 12) pale stripes on a black ground colour. Vertebral stripe greyish brown to brownish white from eye to base of tail. Dorsal stripe is similarly coloured and extends from above eye, coalescing with vertebral stripe on tail. Remaining stripes consist of: brownish white dorsolateral, upper lateral of similar colour, prominent white midlateral and white ventrolateral. Head and tail brown. Limbs brown streaked with black. Belly white. Adult SVL 65 mm.

GENERAL

A skink that prefers to forage close to the protective cover of vegetation, where it quickly darts down a shallow burrow when disturbed, or uses the shelter provided by the vegetation to avoid detection. Also shelters beneath logs, scattered rubbish and inside bulldozer spoil-heaps.

Red-legged ctenotus
Ctenotus labillardieri (Duméril & Bibron 1839)

Stoneville

HABITAT — In this region it is confined to the Darling Range, where it inhabits heathlands, woodlands and rock outcrops. Often found in large numbers on isolated granite outcrops. To the south, at Lake Clifton, it occurs on limestone, and at Lake Mealup on white sand beneath banksia.

DESCRIPTION — Ground colour brown or olive and devoid of pattern on back. Narrow black laterodorsal stripe and distinct white dorsolateral stripe extend from above eye to above hindlimb, becoming wider and diffuse on tail. Upper lateral zone black, with or without a few indistinct pale spots from snout to tail. Midlateral stripe is sharp and white, extending from upper lip to hindlimb and onto side of tail. Similar to *C. delli*, but differs in having reddish limbs boldly marbled with black, rather than brown limbs peppered with black. Belly yellowish. Adult SVL 76 mm.

Lake Mealup

GENERAL Found beneath rocks and logs, usually in an excavated shallow burrow, occasionally inside dead blackboys. Favours granite outcrops, where it is not uncommon to find several individuals beneath the same rock. Unlike other *Ctenotus* species, which are quick to flee when disturbed, in some areas these lizards show little fear of humans and can be easily approached. This skink is restricted to south-west Western Australia, where it is found in wet karri forests, a habitat that other *Ctenotus* species rarely exploit.

Western limestone ctenotus
Ctenotus lesueurii (Duméril & Bibron 1839)

City Beach

HABITAT Found on coastal dunes in limestone areas supporting heath. On the coastal sandplain it is usually associated with banksia and eucalypt woodlands. Absent from the Darling Range.

DESCRIPTION Much like *C. fallens*, from which it can be distinguished because its back pattern usually includes six narrow white stripes rather than four. A large ctenotus with complex and prominent pattern. Ground colour olive-brown, pale-brown to greyish brown in adults, coppery brown in juveniles. A narrow black vertebral stripe extends from nape to base of tail. A black laterodorsal stripe also extends well onto tail, it is edged with white on inner edge. A white dorsolateral stripe extends from behind eye to middle of tail. Upper lateral zone black, with a series of white spots or dashes that may coalesce with white midlateral stripe. The midlateral stripe is broken between upper lip and behind forelimb to form a series of obliquely vertical dark-edged bars. Belly white. Adult SVL 100 mm.

GENERAL Seeks concealment in burrows at the base of shrubs, beneath rocks or rubbish. Favours areas with deep white sand where its close relative *C. fallens* tends to be less common.

Western slender bluetongue

Cyclodomorphus branchialis (Günther 1867)

Genus *Cyclodomorphus*
Fitzinger 1843

An Australian endemic genus containing eight species, of which one is found locally. They are small to moderately large smooth-scaled skinks with very elongate body and short somewhat stumpy limbs, all with five digits. Ear aperture lacks lobules. Parietal separated by interparietal.

A taxonomic revision of members of this genus is currently being undertaken by Glen Shea. It is likely the locally found member will be removed from the synonymy of *Cyclodomorphus branchialis* and defined as a separate species.

Lancelin

HABITAT	Uncommon. Mainly associated with coastal dunes and limestone north of the Swan River, particularly on dunes supporting Beach Spinifex (*Spinifex longifolius*). Occasionally found on pale sands supporting shrublands and woodlands. Absent in the Darling Range.
DESCRIPTION	A moderately large, smooth-scaled terrestrial skink with short limbs, each bearing five digits. The lower eyelid is movable and the ear-opening is small. Ground colour pale-grey, brown to white. Each scale has a small dark anterior spot or 1–3 short dark dashes. Juveniles have darker ground colour and pale spots on head. Belly white. Adult SVL 95 mm.
GENERAL	A nocturnal lizard, although probably also forages during the early morning and at dusk. Feeds on a variety of insects, sometimes consuming snails and small lizards. Shelters beneath dense low vegetation, leaf-litter, limestone, rubbish and inside bulldozer spoil-heaps. When encountered beneath cover, it is usually partially submerged in loose sand. Bears live young; 2–5 in a litter. A birth of four offspring was recorded in late January. The mother was 95 mm SVL and weighed 12.1 g immediately after giving birth. The offspring had an average SVL of 42 mm and weight of 1.18 g.

King's skink
Egernia kingii (Gray 1838)

Genus *Egernia*
Gray 1838

A large genus containing more than 25 species in Australia, with one extending to New Guinea. Four species are found locally. They are medium-sized to very large solid-bodied skinks with well-developed limbs, each bearing five digits. Body scales smooth to strongly keeled. Ear aperture large with one or more lobules. Parietals separated by interparietal. Some species are very inquisitive towards humans.

Adult — Caversham

HABITAT — Associated with limestone formations, rivers and major creek systems on the Swan Coastal Plain. In the Darling Range it may be found on granite outcrops and along streams and water catchment areas. Very common on many of the offshore islands, including Rottnest and Garden.

DESCRIPTION — A large diurnal lizard with stout build and well-developed limbs. Dorsal scales have strong blunt keels. Ground colour brown, greyish brown, greenish brown, dark-grey or black. Juveniles often marked with distinct cream or yellow dashes and spots. This pattern often fades with maturity, although in some populations it is present in adults. Belly cream to grey with darker streaks, mainly beneath throat and tail. Adult SVL 24 cm.

GENERAL — Usually found living in colony groups among limestone and granite. Shelters beneath rocks, in crevices and burrows. On the coast is often seen foraging in coastal dune vegetation, but is nearly always close to rocks for retreat. Mainly a coastal lizard; rarely encountered far inland. Feeds on insects, small lizards, vegetation and the eggs of nesting seabirds. In this regard is useful in the control of seagulls. Bears live young, usually two in a litter.

Juvenile　　　　　　　　　　　　　　　　　　　Kelmscott

Key to *Egernia* species

1	Dorsal scales smooth	**Western glossy swamp egernia** (*luctuosa*)
	Dorsal scales keeled	2
2	Nasal scale strongly grooved behind nostril, back pattern lacks distinct vertebral stripe	3
	No groove behind nostril, back pattern includes distinct vertebral stripe	**South-western spectacled rock egernia** (*pulchra*)
3	Belly grey, upper lips dark (at most spotted with white)	**King's skink** (*kingii*)
	Belly yellow to pinkish, upper lips obviously white	**South-western crevice egernia** (*napoleonis*)

Western glossy swamp egernia
Egernia luctuosa (Peters 1866)

Herdsman Lake

HABITAT	Restricted to dense vegetation surrounding lakes, swamps and rivers. Locally common at Bibra and Herdsman Lake and other large wetlands such as along the Swan River at Maylands. The Perth region is at the northern limit of its geographic range.
DESCRIPTION	A medium-sized terrestrial lizard with smooth glossy scales. Ground colour varies from greenish yellow, yellowish brown to dark brown. Six series of black oblong spots or blotches extend longitudinally from neck to tail. The head scales are unevenly edged with black. Mid and lower lateral zone heavily marked with yellow or yellowish brown flecks. Belly white to yellow, often spotted with black. Adult SVL 13 cm.
GENERAL	A shy secretive skink, rarely found far from the cover of dense vegetation or logs. If disturbed it will dive into water and swim with ease. Favourite basking sites include flattened clumps of bulrushes and horizontal paperbark logs. Becomes nocturnal during hot weather. Around Perth this lizard may be threatened because of the destruction of wetland habitats. Feeds on invertebrates. Produces live young.

LIZARDS

South-western crevice egernia
Egernia napoleonis (Gray 1839)

Ellenbrook

HABITAT
Common in the Darling Range on granite and laterite outcrops, as well as in woodland areas away from rock. On the Swan Coastal Plain it generally inhabits banksia woodlands and coastal heathlands with emergent blackboy (*Xanthorrhoea*). Occurs on Rottnest Island where it co-exists with *E. kingii* on limestone.

DESCRIPTION
A medium-sized lizard with strongly keeled dorsal scales. Ground colour dark-olive or greyish brown with black spots on back tending to align in three longitudinal series. When viewed from the side, a distinct pale area commences on the snout and passes under the eye and through the ear to the forelimb. Scales above mouth white to pale brown, paler than ground colour. A number of small white dots may be scattered over back and sides. Belly salmon-pink to orange-brown. Adult SVL up to 13 cm.

GENERAL
Diurnal and arboreal. Feeds on a variety of invertebrates and will also scavenge scraps at camp sites. Shelters inside dead blackboys and logs, beneath loose bark on trees, under rocks and in crevices. The keeled scales make it particularly difficult for a predator to dislodge. When disturbed it will take cover but will often quickly show itself again, peering inquisitively over a boulder or log. Like most members of the genus *Egernia*, it will colonise in family groups. Produces 2–4 live young in a litter.

South-western spectacled rock egernia *Egernia pulchra* Werner 1910 subsp. *pulchra*

Waroona

HABITAT — Recorded outside the Perth area at Julimar State Forest, near Toodyay. Included in this book as it may occur in scattered eucalypt woodlands over heath, or in similar habitats, in the Darling Range. More common in the humid south-west corner of Western Australia.

DESCRIPTION — A medium-sized skink with weakly keeled dorsal scales. The limbs are relatively short, each with five digits. Ground colour pale yellowish brown to reddish brown with or without a broad vertebral stripe of similar colour extending from neck to base of tail. This stripe is edged by a broad dark dorsal stripe, enclosing a series of pale dorsal spots. Flanks sometimes spotted or densely flecked with cream and dark-brown or black. Eyelids distinctly edged with cream. Lips and ear lobules may be marked with orange. Belly cream to grey. Adult SVL 11 cm.

GENERAL — An attractive lizard that shelters in burrows beneath rocks or fallen timber, or in rock crevices. It is diurnal and terrestrial. Feeds on invertebrates and produces live young.

Broad-banded sandswimmer
Eremiascincus richardsonii (Gray 1845)

Northam

Banded skinks
Genus *Eremiascincus*
Greer 1979

This genus contains two species that are medium-sized, stout-tailed with short well-developed limbs, each with five digits. Widespread throughout arid and semi-arid areas of all mainland States except Victoria. Both species are glossy, each scale with or without a weak keel. Parietals in contact behind interparietal. Lower eyelid moveable. Ear-opening without lobules. One species extends to the eastern limits of this region.

HABITAT Scarce in this region. Only known from a few scattered localities in the Darling Range at Canning Dam and Glen Forrest. Usually on granite. More common to the immediate east of the Range.

DESCRIPTION An attractive medium-sized skink with short well-developed limbs, each with five digits. The lower eyelid is movable and the body scales are very smooth and glossy, with or without a weak keel. Easily distinguished from other similar-sized skinks in this region by having evenly spaced dark-brown to purplish bands: 8–14 between neck and hindlimbs, 19–32 on tail. Bands on the same animal may differ considerably, being oblique, irregular, branching or broken. Ground colour between banding is pale brown to rich golden brown. Belly white. Adult SVL 11 cm.

GENERAL A terrestrial lizard that shelters by day beneath rocks and leaf-litter, becoming active at night to feed on invertebrates like termites. It is well adapted to living in semi-arid to arid areas. The common name for this lizard comes from its behaviour when eluding capture. It dives into loose sand and moves through it as easily as if it were water. Can move very quickly across the ground, using a combination of limbs and lateral body movement. Lays clutches of 3–7 eggs.

REPTILES AND FROGS OF THE PERTH REGION

Southern five-toed earless skink *Hemiergis initialis* (Werner 1910) subsp. initialis

Earless skinks
Genus *Hemiergis*
Wagler 1830

A group of small slender skinks represented by five species found across southern Australia. Absent from Tasmania. Limbs short, with 2–5 short digits. Scales smooth and glossy. Lower eyelid moveable with transparent disc. Ear-opening usually absent and represented only by a slight depression. Supranasals absent, nasals small to moderate, usually separated. Prefrontals usually absent; if present, form a median suture. Usually found beneath cover in a moist habitat. Nocturnal. Two species found locally.

Dwellingup

HABITAT	Found in the Darling Range wherever moist conditions prevail. Prefers laterite soils supporting jarrah woodlands and granite outcrops. Absent on the Swan Coastal Plain.
DESCRIPTION	A small short-tailed skink with five fingers and toes. The lower eyelid is movable, with a transparent disc. External ear-openings are absent, the ear being represented by a depression. Scales are very smooth and glossy. Ground colour reddish brown to dark coppery brown. When present, pattern on back usually consists of 2–4 distinct to obscure longitudinal rows of dark spots from neck to tail. Sides of body greyish, with or without faint dark spots. Belly orange to red from chest to base of tail, remainder of ventral surface grey to bluish grey. Adult SVL 47 mm.
GENERAL	Often occurs in high densities, sheltering beneath any cover, from leaf-litter, rocks and logs to rubbish. A common inhabitant of gardens, where rockeries and compost heaps provide good hiding places. It is nocturnal, although it may forage beneath leaf-litter and other debris during the day. Produces live young, two or three in a litter, measuring 18–19 mm SVL and weighing 0.08–0.09 g (Bush, 1992).

Two-toed earless skink
Hemiergis quadrilineata (Duméril & Bibron 1839)

Cottesloe

HABITAT Occurs in high densities on sandy soils of coastal dunes that support low heath. Moderately common on Swan Coastal Plain in areas supporting banksias and eucalypts and around wetlands, particularly lakes. Absent in Darling Range. Occurs on Rottnest and Garden Islands.

DESCRIPTION A moderately long-tailed semi-burrowing skink with two fingers and toes. Ground colour pale to dark reddish brown, yellowish brown or greyish brown, sometimes with an olive tinge. Back usually has two lines of spots that may coalesce to form a dark stripe extending from neck onto tail. A dark stripe on each side extends along the full length of body and tail. Head, tail and limbs are brownish grey, usually peppered with dark pigment. Belly bright yellow from chest to base of tail, each scale usually with dark edges forming a faint reticulum. Chin and throat whitish, greyish below tail. Adult SVL 70 mm.

GENERAL Seeks shelter beneath leaf-litter, fallen shrubbery, rocks, logs, dead blackboys and rubbish. Often abundant in gardens, particularly in the northern suburbs. A well-shaded and watered yard creates an ideal moist environment for this lizard. Most active at night. Live-bearing, producing 2–5 young in a litter, measuring 25–28 mm SVL and weighing 0.26–0.3 g (Bush, 1992).

Key to *Lerista* species

1	Forelimbs absent, hindlimbs represented by a stump	**Western worm lerista** (*praepedita*)
	Forelimbs present, hindlimbs include two or more toes	2
2	Forelimbs represented by a stump, two toes	**West coast line-spotted lerista** (*lineopunctulata*)
	Forelimb includes two or more fingers, hindlimb with three or more toes	3
3	Two fingers, three toes	**Perth lined lerista** (*lineata*)
	Four fingers, four toes	4
4	Nasals separated or in very short contact	**South-western four-toed lerista** (*distinguenda*)
	Nasals in medium to long contact	5
5	Back pattern consists of black paravertebral spots	**West coast four-toed lerista** (*elegans*)
	Back pattern consists of black paravertebral stripes	**Bold-striped four-toed lerista** (*christinae*)

LIZARDS

Bold-striped four-toed lerista
Lerista christinae Storr 1979

Ellenbrook

Sandswimming skinks
Genus *Lerista*
Bell 1883

Australia's second largest lizard genus, with more than 60 species described to date. A group of elongate smooth-scaled burrowing or cryptozoic lizards. Those species with fingers never have more fingers than toes. Most species have an overshot top jaw, which helps in moving through loose soil. Lower eyelids either movable with a transparent disc or fused, forming an immovable spectacle. All are egg-layers, producing small clutches. They feed on small invertebrates, their eggs and larvae.

HABITAT In this region, it is only known from Rottnest Island and one site at Ellenbrook, where it occurs on pale greyish sands supporting heath with scattered eucalypts. It is more common north of Perth between the Moore River and Badgingarra.

DESCRIPTION A slender skink boldly marked with four black stripes, but also includes two less bold lower lateral stripes. Eyelid immovable and fused to form a transparent spectacle. Ear-opening minute. Four well-developed limbs, each with four digits. Ground colour silver-grey to almost white. Two black paravertebral stripes extend from nape to base of tail. There is an upper lateral stripe passing through eye along body to base of tail and a narrow black lower lateral stripe. Individuals from Ellenbrook are distinctly silver-grey on back between upper lateral stripes and whitish on flanks below these stripes. Tail reddish. Adult SVL 40 mm.

GENERAL Little is known about this species. Those found at Ellenbrook were collected in pit-traps.

South-western four-toed lerista
Lerista distinguenda (Werner 1910)

Stoneville

HABITAT	Common in the Darling Range in most habitats including granite outcrops. Absent on the Swan Coastal Plain.
DESCRIPTION	A small fossorial skink with immovable lower eyelid and four fingers and toes. Nasals separated or in very short contact. Ground colour pale olive-grey, olive-brown to silver-brown, usually with two or four longitudinal series of dark dots running along back from neck to tail. Head is finely specked with black. Broad black upper lateral stripe from side of head to base of tail is bordered below by a narrow white stripe. Tail usually has reddish or orange tinge, brightest in juveniles. Belly white to pinkish, dotted with dark brown under tail. Adult SVL 45 mm.
GENERAL	Although occasionally found beneath leaf-litter, it is most often located in dark greyish sands beneath logs and rocks. Will forage on the surface during the day, particularly in the early morning and late afternoon.

LIZARDS

West coast four-toed lerista
Lerista elegans (Gray 1845)

Canning Vale

HABITAT — Common on the Swan Coastal Plain in most habitat types, favouring sandy soils supporting low heath and woodland, including banksia and eucalypt. Absent from the Darling Range. Occurs on Rottnest Island.

DESCRIPTION — Almost indistinguishable from *L. distinguenda* but for the medium to long contact of the nasal scales. Eyelid immovable. Four fingers and toes. Ground colour olive-brown to greyish. A paravertebral series of dashes extends from neck to tail. A wide dark-brown upper lateral stripe edged below with white extends from face, along body and becomes obscure on sides of tail. Top of head unpatterned and darker than back. Like *L. distinguenda*, the reddish coloration on tail is brightest in juveniles. Belly whitish, sometimes clouded with grey. Below tail usually dotted with greyish brown. Adult SVL 43 mm.

GENERAL — Shelters beneath limestone rocks, leaf-litter and logs. Also uses disturbed areas such as spoil-heaps and loose soil beneath rubbish, particularly corrugated iron. A surface-dweller and burrower. This lizard is often observed darting among low vegetation during the day. It has been recorded aggregating in a suburban backyard (Browne-Cooper, 1992).

Perth lined lerista
Lerista lineata Bell 1833

Port Kennedy

HABITAT

Found in southern suburbs and dunes of the Swan Coastal Plain. Restricted to pale sands supporting heathlands and shrublands, particularly in association with banksias. Also occurs on Rottnest and Garden Islands.

DESCRIPTION

A small slender lizard with a slightly protrusive snout and immovable lower eyelids. Forelimbs small with two fingers, hindlimbs relatively long with three toes. Ground colour pale brownish grey, being darker and browner along centre of back between paravertebral lines. Prominently striped with a wide black upper lateral stripe and narrow greyish white midlateral stripe. Head and lips blotched with black. Belly dotted or flushed with grey. Adult SVL 55 mm.

GENERAL

Shelters in leaf-litter and upper layers of loose sand at bases of shrubs, inside bulldozer spoil-heaps and abandoned stick-ant nests. Occasionally found in loose soil beneath discarded fibro and corrugated iron. The geographic range of this lizard is very restricted, extending from the southern suburbs of Perth to Mandurah. It was recognised as threatened, because of urban development, until its removal from the gazetted list in 1990. Much of its remaining habitat is backyard gardens.

West coast line-spotted lerista
Lerista lineopunctulata (Duméril & Bibron 1839)

Trigg

HABITAT — Common on coastal dunes supporting heath in northern suburbs. Also occurs on sandplains in banksia and eucalypt woodland. Absent from Darling Range. Occurs on Rottnest Island.

DESCRIPTION — A large and robust lerista with immovable lower eyelids. Forelimbs are represented by tiny clawless stumps and hindlimbs are small, bearing two clawed digits. Ground colour pale-grey to greyish brown with six or more longitudinally aligned dots or dashes from neck to tail. These may be obscure or distinct. Occasionally markings are transversely elongate or squarish. Lips have dark bars. Belly whitish. Adult SVL 100 mm.

GENERAL — Inhabits top layer of loose sand beneath leaf-litter at bases of trees and shrubs, also beneath rocks and logs. The build and shape of this lizard make it an accomplished sand swimmer. It regularly wriggles considerable distances between leaf-litter deposits during night or day. The majority of small snake-like tracks seen on local northern beaches and dunes are caused by this lizard moving beneath the surface.

Western worm lerista
Lerista praepedita (Boulenger 1887)

Melaleuca Park

HABITAT	Common on coastal dunes and pale sands supporting heath and woodlands. More abundant north of the Swan River than to the south. Absent from the Darling Range. Occurs on Rottnest and Garden Island.
DESCRIPTION	A very slender, almost limbless skink with movable lower eyelids and protrusive snout. Forelimbs completely absent, hindlimbs reduced to stumps. Ground colour very pale olive-grey to pale silver-brown with two paravertebral series of blackish brown dashes forming distinct or obscure lines from neck to tail. A broad blackish brown upper lateral stripe extends from head, through eye to tip of tail. Belly greyish white with each scale edged anteriorly with blackish brown. Adult SVL 65 mm.
GENERAL	Shelters in loose soil beneath leaf-litter at bases of coastal shrubs, beneath limestone rocks, logs, stumps and in sand beneath rubbish.

Common dwarf skink
Menetia greyii Gray 1845

Genus *Menetia*
Gray 1845

An Australian endemic genus of very small skinks with four fingers and five toes. Lower eyelid immovable; instead fused to form a transparent spectacle.

Northam

HABITAT	Found throughout the Perth region in most habitat types, including disturbed areas such as backyards.
DESCRIPTION	A very small smooth-scaled terrestrial skink with moderately well-developed limbs bearing four fingers and five toes. Lower eyelid is immovable and fused to form a transparent spectacle. Ear-openings are minute. Grey to greyish brown, usually with black flecks arranged longitudinally along back. A black upper lateral stripe bordered below by white; most distinct anteriorly. Some populations have bright-yellow bellies, whereas others are silver-grey. Males during breeding have reddish chests and forelimbs. Adult SVL up to 38 mm.
GENERAL	One of Australia's most abundant and widespread lizards. A diurnal forager that feeds on small invertebrates. Lays two clutches of two or three eggs each year. Eggs measuring 7–8 mm long and 4–5 mm wide. Depending on incubation temperature, these take 45–104 days to hatch. Hatchlings SVL 12–16 mm and weight 0.03–0.08 g (Bush, 1983b).

Western pale-flecked morethia
Morethia lineoocellata (Duméril & Bibron 1839)

Genus *Morethia*
Gray 1845

An Australian endemic genus of small skinks with five fingers and toes. Lower eyelid immovable; instead fused to form a transparent spectacle. Parietal and frontoparietals fused into a single large scale.

Breeding male   Lake Mealup

HABITAT — On the coast, it is found in limestone formations and coastal dunes supporting low heath. Further inland it occurs in banksia and eucalypt woodland. Rarely recorded in the Darling Range. Occurs on Rottnest and Garden Island.

DESCRIPTION — Similar to *M. obscura* but usually has bolder pattern and supranasals fused to nasals. A small skink with smooth shiny scales and five fingers and toes. Eyes relatively large with immovable eyelids, each bearing a transparent spectacle. Ground colour olive-grey, olive-brown to coppery brown with distinct white-edged black spots (ocelli) and black flecks over body and limbs. Pale narrow dorsolateral stripe may extend from neck to base of tail. A white midlateral stripe, usually well-defined, extends from upper lip to hind limb, and is bordered by a narrow dark lower lateral stripe. Belly white. During breeding season, males develop a brilliant reddish flush on lower lip and throat. Adult SVL 49 mm.

GENERAL — An agile sun-loving lizard that forages during the day beneath low vegetation. When disturbed it quickly darts between shrubs and remains motionless, merging in with the foliage. During cool weather it is often dug up in sand beneath rocks, logs and rubbish. Feeds on small invertebrates. Lays as many as five eggs in a clutch.

Southern pale-flecked morethia
Morethia obscura Storr 1973

Breeding male — Mount Dale

HABITAT

Occurs in most habitat types in this region, from heath and shrublands on coastal dunes to eucalypt and banksia woodlands on Swan Coastal Plain. Favours wetter and more heavily vegetated areas than *M. lineoocellata*. Abundant in the Darling Range in shrublands and jarrah woodlands and vegetation fringing granite outcrops. Occurs on Garden Island.

DESCRIPTION

Similar to *M. lineoocellata* but differs in attaining a greater size, more obscured pattern and having distinct supranasals. Adult SVL 55 mm. Ground colour grey to olive-brown. Pattern is weak to absent, consisting of indistinct white-centred black ocelli or dark flecks, most distinct on posterior part of body and tail. Indistinct blackish brown upper lateral zone extends from behind eye to base of tail. Rarely any indication of pale dorsolateral stripe. Top of head coppery brown. Belly white. In spring, breeding males have reddish throats.

GENERAL

An active forager in leaf-litter at the bases of trees and shrubs and around granite and laterite rocks. Often found in very disturbed habitats. Lays 1–5 (usually four) eggs in a clutch, measuring 9–10 mm long and 4.5–6 mm wide. Incubation periods of 29 days at 30°C and 95 days at an estimated temperature of 19.6°C have been recorded. Hatchlings measured 19 mm and weighed 0.15 g (Bush, 1992).

Western bluetongue
Tiliqua occipitalis (Peters 1863)

Blue-tongued skinks
Genus *Tiliqua*
Gray 1825

Very large solidly-built skinks with short limbs, all with five digits. Lower eyelid moveable without transparent spectacle. Includes Australia's best-known lizards.

Ellenbrook

HABITAT — In this region, associated with coastal and near-coastal dunes supporting heath. Locally common in the Rockingham-Port Kennedy area, near Bullsbrook and north of Wanneroo. Scarce elsewhere within the Perth area.

DESCRIPTION — A large terrestrial smooth-scaled skink with a long, slightly depressed head and longer more slender tail than the bobtail. Ground colour pale yellowish brown to light greyish brown, marked with 7–11 broad dark bands across body and tail. Pale scales between bands are usually dark-edged and these pale areas may enclose additional narrow incomplete bands. A broad dark stripe extends from eye to about ear. Belly cream to white, sometimes having an obscure extension of bands laterally that merge slightly with belly coloration. Adult SVL 27 cm.

GENERAL — An easily identifiable diurnal lizard. Often found moving across tracks or over dunes on the coast. When harassed, it flattens its body, opens its mouth to display a pink interior with large blue tongue and hisses as it inflates and deflates rapidly. Shelters in rabbit burrows or excavates a shallow burrow beneath rocks and dead vegetation. Omnivorous, feeding on a wide range of invertebrates, gastropods (snails and slugs), carrion, flowers and fruit. Produces litters of 4–10 live young in late summer, measuring SVL 10–12 cm and weighing 25–30 g (Bush, 1992).

LIZARDS

Shingleback, bobtail or sleepy lizard *Tiliqua rugosa* (Gray 1825)

Mainland subsp. *rugosa* — Langford

HABITAT

Common throughout this region in most habitats, although less common around swamplands and wetlands. Still occurs within many inner suburbs and pockets of bushland. Mainland subspecies also found on Garden Island (Robinson *et al.*, 1987). A separate subspecies, *Tiliqua rugosa konowi*, occurs on Rottnest Island. It is smaller and darker than the local mainland population.

DESCRIPTION

A large skink with broad triangular head and short blunt tail. Scales are large, irregularly shaped and rugose. Ground colour varies, from pale to dark brown, olive, grey or black. Back and tail are usually marked with pale spots, blotches or streaks, which sometimes form obscure to distinct cross bands. Head usually pale orange-brown. Belly white or off-white with irregular greyish brown stripes, bands or blotches. Adult SVL 30 cm.
(Continued on next page)

Shingleback, bobtail or sleepy lizard *(Continued)*

Rottnest subsp. *konowi* (Mertens 1958)   Rottnest Is.

GENERAL

The best-known and most easily recognised Australian lizard. The large number of different common names around the country indicates the familiarity Australians have with this species. During the warmer months many bobtails fall victim to vehicles on roads, unfortunately some motorists intentionally hit these placid animals. Like the Western Bluetongue (*Tiliqua occipitalis*), it will attempt to deter harassment by facing the molester with mouth agape, hissing and displaying the large blue tongue. It has powerful jaws with which it can inflict a painful bite (fingers and dogs' noses beware!). Shelters beneath dead vegetation and rubbish such as corrugated iron, and in burrows. Omnivorous, feeding on a wide range of invertebrates, flowers and fruits, also garden slugs and snails. Produces one or two (rarely three) live young, which at birth are a third to half adult length.

Snakes

(Order Squamata, suborder Serpentes)

THE lizards and snakes have been treated as a single group, and you will find an introduction to the group on page 54.

Key to snake families

1	Eyes inconspicuous, body worm-like without tapering tail	**blind snake** (Typhlopidae)
	Eyes conspicuous, body not worm-like, tail obviously tapering or laterally compressed	2

2	Tail laterally compressed into paddle shape	**sea snake** (Hydrophiidae)
	Tail obviously tapered	3

3	Midbody scale rows 39 or more. Cloacal spurs present	**python** (Boidae)
	Midbody scale rows 23 or fewer. Cloacal spurs absent	**fixed front-fanged land snake** (Elapidae)

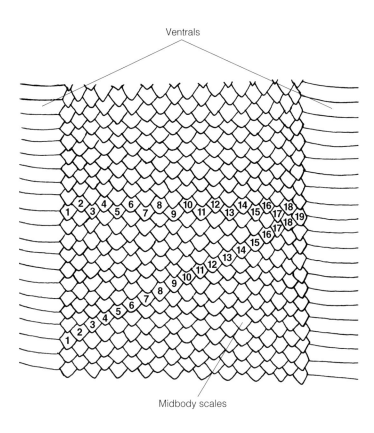

Method of counting midbody scales of snakes

LAND SNAKES

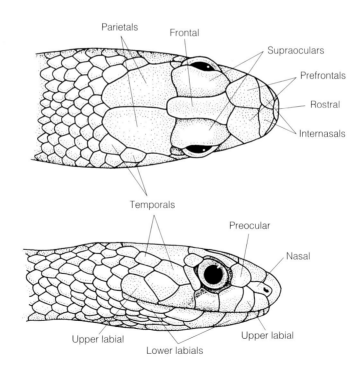

Head scales of snakes

Key to the land snake genera and species

This key will allow identification to the specific level in the family Elapidae, and to generic level in the pythons (Boidae) and blind snakes (Typhlopidae). A further key has been included for the blind snakes. (See pages 144, 145, 148 for methods of counting scales.)

1	Body worm-like, tail without taper, eyes indistinct (each represented by a dark spot beneath head scales)	**blind snake** (see p. 151) (*Ramphotyphlops* species)
	Body snake-like, tail tapering, eyes distinct	2
2	Body stout, tail noticeably distinct from body and includes grub-like appendage on end	**Common death adder** (*Acanthophis antarcticus*)
	Body slender to moderately stout, tail with gradual taper (hardly distinct from body), no grub-like appendage	3
3	Midbody scales in 23 or fewer rows	4
	Midbody scales in 39 or more rows	**python** (*Morelia* or *Antaresia* species)
4	Midbody scales in 15 rows	5
	Midbody scales in 17 rows	11
5	Scales beneath tail divided (in two rows)	6
	Scales beneath tail undivided (in one row)	9
6	Tail long, 35 or more scales along underside of tail	**Reticulated whip snake** (*Demansia psammophis*)
	Tail short, 30 or fewer scales along underside of tail	7

7	Body and tail banded	**Banded sand snake** or **Jan's banded snake** (*Simoselaps bertholdi*)
	Body and tail not banded (two black bars on head only)	8

8	Dark vertebral stripe usually present	**Black-striped snake** (*Neelaps calonotus*)
	Dark vertebral stripe absent	**Black-naped snake** (*Neelaps bimaculatus*)

9	Head shiny black, no transverse bar on neck	10
	Head greyish, black transverse bar on neck	**Crowned snake** (*Drysdalia coronata*)

10	Black on head extends along back as broad dark vertebral zone	**Black-backed snake** (*Rhinoplocephalus nigriceps*)
	Black on head extends no further than neck, no dark vertebral zone	**Gould's hooded snake** (*Rhinoplocephalus gouldii*)

11	Scales beneath tail all divided or all undivided	12
	Scales beneath tail undivided anteriorly, divided posteriorly	**Mulga snake** (*Pseudechis australis*)

12	Scales beneath tail undivided	13
	Scales beneath tail divided	14

13	Belly cream, yellow or orange anteriorly, dark-grey posteriorly	**Western tiger snake** (*Notechis scutatus*)
	Belly olive to pink for full length	**Bardick** (*Echiopsis curta*)

Key to the land snake genera and species (continued)

14 Eye small (diameter no greater than distance between eye and mouth)   15

     Eye large (diameter greater than distance between eye and mouth)   16

15 Body whitish with numerous red spots between narrow black bands, interspaces much wider than bands   **Narrow-banded snake** (*Simoselaps fasciolatus*)

     Body brown between dark-brown bands, interspaces and bands equal in width   **Southern half-girdled snake** (*Simoselaps semifasciatus*)

16 Midbody scales in 17 rows   **Gwardar** or **Western brown snake** (*Pseudonaja nuchalis*)

     Midbody scales in 19 rows   **Dugite** or **Spotted brown snake** (*Pseudonaja affinis*)

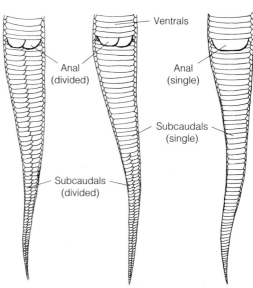

Arrangement of scales below snake tail

Blind or worm snakes
(Family Typhlopidae)

A large widespread family with roughly 150 species in two genera worldwide. In Australia the genus *Ramphotyphlops* Fitzinger 1843 is represented by about 40 species, of which three occur in the Perth region. Many of Australia's blind snakes are known only from a single specimen or from small isolated localities. Because of their subterranean habits they have attracted little interest from herpetologists, and consequently the ecology of many species is poorly understood.

They are slender to robust worm-like snakes. The mouth is small and set beneath an overshot top jaw. The tail is short and terminates in a short conical spine, which assists the snake in its movement through the soil. The eyes are reduced to a pair of dark spots covered by ocular scales, and can probably only distinguish light from dark. The body scales are very smooth, close-fitting and well-adapted for resisting ant stings. The belly or ventral scales are the same size as those around the body.

Blind snakes are fossorial, spending most of their lives underground. They shelter in ant nests, termite mounds or in soil beneath leaf-litter, rocks and logs. Occasionally one may be unearthed in the garden or encountered foraging on the surface at night, particularly after rain in the warmer months. All are oviparous, producing anywhere between two and 25 eggs in a clutch. They apparently feed exclusively on ant eggs, larvae and pupae and termites, which are usually eaten below the surface during raids on nests.

All blind snakes are nonvenomous and quite harmless. To deter predators they use an unpleasant odour, which is released from well-developed anal glands. This odour may also have a role in reproduction, assisting snakes in locating mates.

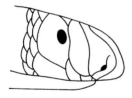
Ramphotyphlops australis

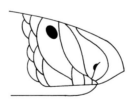
Ramphotyphlops pinguis

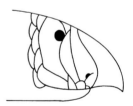
Ramphotyphlops waitii

Lateral head views of local blind snakes

Key to blind snake (*Ramphotyphlops*) species
(See page 144 for method of counting midbody scales.)

| 1 | 20 scale rows at midbody | 2 |
| | 22 scale rows at midbody | **Southern blind snake** (*australis*) |

| 2 | Body stout, fewer than 350 ventrals | **Fat blind snake** (*pinguis*) |
| | Body slender, more than 500 ventrals | **Beaked blind snake** (*waitii*) |

Southern blind snake
Ramphotyphlops australis (Gray 1845)

Gravid female   Sullivan Rock

HABITAT — Common throughout the Perth region, favouring damper conditions in the Darling Range. May be found in all habitats as well as suburban gardens. Occurs on Rottnest Island.

DESCRIPTION — A moderately large and robust blind snake, with snout rounded from above and rounded to weakly angular in profile. Midbody scales in 22 rows, 270–357 ventrals, 10–20 subcaudals. Ground colour purplish brown to purplish grey or pink in adults, and pink in juveniles. Belly cream. Usually a distinct jagged colour junction formed between belly and ground colour. Total length 45 cm.

Mount Helena

GENERAL   The most common blind snake. Shelters beneath a wide variety of cover, from rocks and logs to old building materials. Houses adjacent to bushland often have these snakes inhabiting the gardens. The first impression after unearthing a blind snake is of an oversized worm, though closer examination of the head reveals a flickering forked tongue, which identifies it as a snake. A clutch of five eggs was laid in late February: length 20–25 mm, width 5–7 mm, weight 0.24–0.3 g. These eggs hatched after 55–59 days at 29°C. Neonate total length 90–98 mm, weight 0.5 g (Maryan, 1988). Nonvenomous.

Fat blind snake
Ramphotyphlops pinguis (Waite 1897)

York

HABITAT	Found throughout the Darling Range in woodlands and rocky outcrops. Absent from the Swan Coastal Plain.
DESCRIPTION	A large and very stout blind snake with a weakly trilobed snout when viewed from above, and slightly angular profile. Midbody scales in 20 rows, 230–270 ventrals, 10–20 subcaudals. Ground colour light-grey, dark-grey to greyish brown. A pale smudge may be present on each scale, which align to form a longitudinal series of spots on anterior portion of body. Belly whitish, either contrasting sharply or weakly with ground colour. Total length up to 45 cm.
GENERAL	This is the thickest blind snake found locally. It has been found in bull-ant nests. Its thick strong body may enable it to feed on the larvae and pupae of the larger more aggressive ant species without injury from the stinging and biting insects. Nonvenomous.

Beaked blind snake
Ramphotyphlops waitii (Boulenger 1895)

Maddington

HABITAT
Found throughout the Darling Range and adjoining coastal plain south to about Armadale in this region. Most records come from the vicinity of the Upper Swan, Midland and Guildford areas. Scarce on the Swan Coastal Plain towards the coast.

DESCRIPTION
A long, extremely slender blind snake, with trilobed snout when viewed from above, and distinctly beak-like snout in profile. Midbody scales in 20 rows, ventrals 500–670, subcaudals 10–26. Ground colour yellowish brown, caramel-brown to dark purplish brown. Often paler on snout than body. Belly much paler than body. Total length up to 60 cm.

GENERAL
The very slender body may enable this blind snake to enter the nests of much smaller ants than those fed on by the heavier species. Nonvenomous.

Pythons

(Family Boidae)

A group of small to extremely large nonvenomous snakes ranging in size from the Pygmy Python (*Antaresia perthensis*), with a total length of 50 cm, to an unconfirmed report of an Amethyst or Scrub Python (*Morelia amethistina*) of eight metres long. This is indeed one of Australia's largest snakes, however, the Pilbara Olive Python (*Liasis olivaceus barroni*) has been confirmed to be just under six metres long, and is definitely Western Australia's largest snake.

Two species of python occur in the Perth region; both are medium-sized. One is common, the other is scarce locally. All Australian pythons (except for the genus *Aspidites*) have heat-sensitive pits on the labials or lip scales and a broad head and relatively narrow neck. These pits assist in the locating of warm-blooded prey both by day and night. Body scales are small and numerous. Head scales are either small and fragmented or large and symmetrical.

Pythons are mainly nocturnal, although they can be encountered basking during the day, particularly when digesting a large meal.

Droppings of Carpet Python

Most are excellent climbers, being equally at home in trees, on rock outcrops and on the ground. They generally feed on vertebrates, especially birds and mammals. At least one *Aspidites* species feeds on other reptiles including venomous snakes, hence the lack of need for heat-sensitive pits in that genus. Prey is overcome by constriction with tight coils of the python's body until the victim is unable to breath. Pythons don't crush their prey to death as is popularly believed.

They are the only Australian snakes to have vestigial pelvic bones. They also have cloacal spurs, which are often referred to as vestigial hindlimbs. These are located on either side of the vent. The spurs are larger in males and are used to stimulate the female during courtship.

Pythons are egg-layers, producing 3–30 eggs in a clutch. The eggs often adhere together as a cluster. The female fasts prior to egg-laying and remains coiled around the eggs during incubation. At this time her body temperature is slightly raised by muscular spasms and shivering. Although nonvenomous, pythons are equipped with many sharp, backward-curved teeth, and can inflict a painful bite if molested or roughly handled. Their diet includes vermin and rodents such as rabbits, rats and mice, and they are important in the control of these pests.

SNAKES

Stimson's python
Antaresia stimsoni (Smith 1985) subsp. *stimsoni*

Children's pythons
Genus *Antaresia*
Wells & Wellington 1984

This genus is restricted to Australia and includes the world's smallest python. Prior to a revision by Kluge (1993) members of it had been placed variously in the genera *Liasis* and *Morelia*.

A small (less than 1.2 metres) python with prehensile tail. Head covered with large symmetrical scales, and heat-sensitive pits in some scales on the lips.

Darlington

HABITAT Scarce throughout this region. Perth is the southern limit of its geographic range. Generally restricted to the Darling Range, favouring granite outcrops.

DESCRIPTION A small python with large symmetrical plate-like head scales. Midbody scales in 39–47 rows, 260–300 ventrals, anal scale entire, 40–50 subcaudals, mostly divided. Ground colour pale brown, yellowish brown to whitish. Body randomly marked with darker brown to reddish brown blotches. These blotches may be either elongate or circular in shape. A pale stripe extends along each side of the neck for some distance along body. Head usually blotched above with dark streak from the nostril passing through the eye to the neck. Belly whitish. Adult length up to 100 cm.

GENERAL A nocturnal python, although occasionally encountered basking or foraging during the day. Shelters in hollow tree limbs, rock crevices, termite mounds, animal burrows and under exfoliated (flat layers of) granite slabs. Diet includes small birds, mammals and lizards. When disturbed it may curl up into a tight ball, hiding its head within the coils. Prior to a revision of the small pythons in 1985 by L.A. Smith, it was placed with the Children's Python (*Antaresia childreni*), which is now restricted to northern Australia. Harmless.

Southern carpet python

Morelia spilota (Lacépède 1804) subsp. *imbricata* (Smith 1981)

Carpet and diamond pythons
Genus *Morelia*
Gray 1842

This genus includes Australia's largest snakes and is represented in the Torres Strait Islands and New Guinea. With the exception of one species, generally restricted to northern Australia. All but one Australian species boldly patterned.

Moderately large to very large python (up to six metres) with prehensile tail. Head covered with small fragmented scales or large symmetrical scales, and heat-sensitive pits in some scales on the lips.

Two Rocks

HABITAT — Inhabits substantial undisturbed patches of bushland near Perth, especially catchment areas and rocky outcrops in the Darling Range. Large-scale development on the Swan Coastal Plain has probably reduced its numbers. Moderately common in the north around Neerabup and Yanchep National Parks, where limestone formations encompass heath. Common on Garden Island.

DESCRIPTION — A large python with small fragmented head scales. Midbody scales in 40–50 rows, 240–280 ventrals, anal scale entire, 60–82 subcaudals, mostly divided. Dorsal scales smooth, lanceolate (shaped like a spear-head) and strongly overlapping at rear of body. There are unconfirmed anecdotal reports in Western Australia, claiming individuals of up to four metres long. Ground colour variable, mainly consisting of greenish brown or blackish brown, and marked with numerous irregular dark-edged cream, yellow to pale-brown blotches. These blotches merge to form transverse zones on the dorsum and those on anterior flanks often coalesce forming a stripe. Belly white, cream or yellow, unmarked or with bold black blotches. In the Darling Range, individuals often with less bold dorsal pattern and dark-grey to olive-green with unmarked white belly. On the

Karragullen

Swan Coastal Plain, often with bold contrasting dorsal pattern of green and black, with cream or yellow belly boldly marked with black. Total length up to 250 cm.

GENERAL

Probably Australia's most well-known python. Mainly nocturnal though sometimes seen basking and moving about during day. Arboreal, terrestrial and rock-dwelling. Shelters in a wide range of sites from hollow tree limbs to rock crevices and the burrows of other animals. Feeds mainly on mammals and birds, with the occasional lizard being taken, especially by juveniles. Usually a quiet snake that shows little objection to being handled, although some individuals can be defensive, hissing loudly and striking with open mouth. The combined effect of habitat destruction, predation by foxes and feral cats and bushfires has made a noticeable impact on its numbers. It is currently gazetted as a threatened species. There are only limited reproductive data available on this south-western subspecies. It has been recorded as laying clutches of 14, 17 and 20 eggs measuring 51–63 mm long and 36–45 mm wide and weighing 46–52 g. Hatching occurred after 63–71 days at 30°C. Neonate SVL 24–38 cm, weight 14.5–31 g (Bush, 1988 and unpublished data). This snake is harmless.

Front-fanged venomous land snakes

(Family Elapidae)

A large family containing about 200 species in 50 genera worldwide. Elapids are well represented in Australia, with more than 80 species in 22 genera. Elapids are characterised by having a pair of relatively short, fixed, effectively hollow fangs at the front of the upper jaw. These are enclosed in a fleshy sheath and are connected by a duct to venom glands located on each side of the head behind the eye. When an elapid bites, venom enters the wound under pressure caused by the voluntary contraction of muscles around the venom glands.

Snake venom is a highly specialised saliva. It can immobilise prey and speed up digestion. Some African and Asian elapids have evolved specialised fangs to allow the venom to be used as an effective defence. The Ringhals and Black-necked Cobras have openings at the front of the fangs so that venom can be sprayed forward into the face of an attacker. This causes temporary blindness and confusion, allowing the snake time to escape.

The large venomous elapids are the snakes most commonly encountered in Australia, because these have adapted successfully to our alterations of the environment. The best place to look for brown snakes such as Dugites or Gwardars is in agriculturally developed areas. They might easily be found in a farmer's rubbish dump or an old deserted house.

Elapids occur throughout Australia, with several species represented in any region. They are generally terrestrial, with only a few species considered semi-arboreal or semi-aquatic. All feed on vertebrates. The small species take lizards and frogs, whereas the large species take birds and mammals. Much of their food consists of introduced mice, rats and rabbits, so along with the pythons, elapids play an important part in controlling the pest species.

Reproduction involves egg-laying in some species (oviparous) or live-bearing in others (viviparous). The eggs are leathery or parchment-shelled, whereas the free-born young arrive in a clear membranous sac. The live-bearing species are most abundant in cooler climates; the three species of elapid snake found in Tasmania are viviparous.

In the Perth region this family is represented by 15 species, five of which have potentially dangerous venom.

Common death adder
Acanthophis antarcticus (Shaw & Nodder 1802)

Death adders
Genus *Acanthophis*
Daudin 1803

This genus comprises three described species (but there may be as many undescribed species) in Australia, Indonesia and New Guinea. They are widely distributed throughout Australia except in the extreme south-east.

All the death adders are moderately large thickset snakes with a distinctive triangular head and relatively narrow neck. The tail is slender, terminating in a small soft spine of modified scales, *(Continued opposite)*

Canning Dam

HABITAT	Locally confined to the Darling Range between Mt Helena and Jarrahdale. Found in jarrah woodlands adjacent to granite outcrops and along densely vegetated creeks.
DESCRIPTION	A very stout snake with a broad triangular head, protrusive top jaw, relatively narrow neck and short tail with segmented tip. Dorsal and lateral scales are smooth to slightly keeled. Midbody scales in 21 (rarely 23) rows, 110–125 ventrals. Anal scale undivided. Subcaudals 36–50, undivided anteriorly and divided posteriorly. Dark reddish to greyish brown with numerous pale transverse bands, usually two scales wide. Lips white with dark vertical bars. Belly white with dark flecks. Attains about 75 cm total length.

and usually contrasts with the body colour. The body scales are smooth to strongly keeled and non-glossy. The eyes are small with vertical pupils. It is a unique genus; death adders are unlike any other elapid in Australia. In appearance death adders are like the vipers of Europe and Asia. They are Australia's ecological equivalent to the true vipers. They have large venom glands and fangs that can be rotated further forward and backward than those of any other typical elapid. The venom is primarily neurotoxic, commonly causing paralysis in victims.

Karagullen

GENERAL

Both diurnal and nocturnal. During the day it lies motionless, half buried in leaf-litter or under low vegetation. It is a sluggish snake, which prefers to wait for prey rather than actively forage. While lying camouflaged, the modified tail tip is wriggled, presumably to mimic a grub. This movement attracts a lizard or bird to within range and the snake strikes with lightning speed.

During October and November, adult males are constantly on the move in search of females. Females do not seem to move much until late summer, when increased thermoregulatory behaviour occurs in pregnant individuals. Produces up to 30 live young in a litter. Females attain greater size than males.

The death adder relies on camouflage to remain hidden from predators and is therefore unlikely to move away from an approaching bushwalker. For this reason there is a remote chance of treading on one while walking within its known range. Although **DANGEROUSLY VENOMOUS**, only two fatalities between 1981 and 1994 are believed to have involved the bite of this species. Recently, a Common Death Adder was thought to have bitten a woman at Wilson. The Common Death Adder is unlikely to occur naturally in that suburb. However, more and more small snakes are being inadvertently moved around in firewood.

Reticulated whip snake
Demansia psammophis (Schlegel 1837) subsp. *reticulata* (Gray 1842)

Whip snakes
Genus *Demansia*
Günther 1858

A group of small to medium-sized, slender, long-tailed elapids with large eyes. The diameter of the eyes is markedly greater than the distance from the mouth to the eye. There are 11 described taxa and several undescribed forms occurring widely throughout the continent. In most parts of Australia, at least one species is likely to be encountered. One species extends north to New Guinea.
(Continued opposite)

Sorrento

HABITAT — Uncommon locally. Found on the coastal dunes and on sandplain supporting heath and banksia woodlands. Occurs rarely in the Darling Range. Most likely to be observed north of the Swan River.

DESCRIPTION — A moderately large slender green snake. Smooth matt dorsal scales in 15 rows at midbody, 170–205 ventrals. Anal scale divided, 65–90 subcaudals, all divided. Colour usually a beautiful lime-green, becoming lemon-yellow on flanks. Some individuals are olive-green or dark-green. Each dorsal scale is finely to broadly edged with black along posterior margins. Belly white to greyish. Head markings concentrated around eyes, as in genus notes. Total length to 100 cm.

The dorsal scales are smooth and non-glossy, in 15 rows at midbody. The anal plate and subcaudal scales are divided. The frontal and nasal scales are long and narrow, the nasal being divided by the nostril (except in one species) and in contact with the preocular. The preocular scale is normally separated from the frontal. No subocular scales. In most species a 'comma'-shaped marking encircles the eye, the tail of which extends back towards the angle of the mouth. Occasionally a dark or pale band is present on the neck.

Jandakot

GENERAL

A fast-moving diurnal snake that can be found sheltering beneath rocks, logs and rubbish, but is most often seen on the move during the day. You will usually get only a brief glimpse of this very alert snake. It seems to be intolerant of any disturbance, as indicated by its disappearance from remnant bushland areas such as Kings Park, where it has not been seen for some time. It is venomous but not considered harmful. In humans, a bite will cause mild local symptoms such as swelling. Large individuals should be treated with caution. Feeds almost exclusively on lizards, which are caught on the run. No doubt the large eyes allow it to maintain sight of prey. Although generally diurnal has been observed on roads at night. Oviparous, laying 3–9 eggs in a clutch.

Crowned snake
Drysdalia coronata (Schlegel 1837)

White-lipped snakes
Genus *Drysdalia*
Worrell 1961

Moderately small to very small slender to stout elapids. Eyes large with round pupils. Lips distinctly white. Undivided anal and subcaudal scales. Internasal scales present, suboculars absent. Belly usually brightly coloured. Restricted to the southern parts of the continent. Combat has been observed between male Crowned Snakes (*D. coronata*).

Adult — Canning Vale

HABITAT

Generally scarce in the Perth region, being found mostly south of the Swan River on the Swan Coastal Plain. May occur in the Darling Range. Inhabits wetlands, woodlands and both dry and wet lowlands supporting melaleucas and blackboys. Its northern limit of distribution is this region.

DESCRIPTION

A small, moderately robust snake with smooth matt dorsal scales in 15 rows at midbody. Ventrals 130–153, undivided anal scale, 39–53 subcaudals, all undivided. Ground colour pale brown, reddish brown, olive or green. A black bar crosses the neck and extends forward along each side of the head to meet on the nose. Its common name refers to the black head markings, which resemble a crown. The top of the head is bluish grey in many individuals, contrasting significantly with the ground colour. The lips are white with reddish brown flecks that extend onto the chin. Belly cream to orange-pink. The anterior margin of each ventral scale is much darker than the posterior margin. Adult total length 60 cm.

Juvenile   Canning Vale

GENERAL

Diurnal and nocturnal. Usually found during the day foraging or basking under or on the edge of low thick vegetation and grass tussocks. In this region, it shelters inside blackboy stumps, beneath rubbish, especially pieces of corrugated iron lying on grass, and amongst thick deadfall vegetation. Around wetlands or swamps its food consists of frogs, and in sandy areas it feeds on skinks. Bearing 3–9 live young in March or April, neonate SVL 100–140 mm, weight 1.4–2.5 g (Bush, 1992). Venomous, but not considered harmful.

Bardick
Echiopsis curta (Schlegel 1837)

Genus *Echiopsis*
Fitzinger 1843

A genus restricted to the south of the continent. The two members are medium-sized stout elapids. Internasal scales are present, suboculars absent. The anal and subcaudal scales are undivided.

Guilderton

HABITAT

Uncommon on Swan Coastal Plain and seems to be absent from the Darling Range. Mainly recorded north of the Swan River, inhabiting coastal dune vegetation, heathlands and banksia woodlands.

DESCRIPTION

A small to medium-sized robust snake with relatively large bulbous head, which is distinct from the neck. Eyes large with marginally elliptic pupils. Dorsal scales smooth, matt and in 17–21 rows (usually 19) at midbody. Ventrals 120–144, undivided anal scale, 27–43 undivided subcaudals. Ground colour olive-grey, brown, reddish brown or black. Sides of head and neck often flecked with white, mainly concentrated on lips. A pale streak (often diffuse) may also be present from eye to side of neck. Belly salmon-pink, pale-yellow, cream to greyish white. Attaining an adult total length of 70 cm.

Joondalup

GENERAL

This snake is often mistaken for a Death Adder, although is easily distinguished by its uniform body colour. It is active both day and night and is often observed on roads and tracks on the coast just north of Perth. When found during the day it is usually foraging amongst vegetation on overcast days or basking in diffused sunlight close to cover. Shelters among leaf-litter, beneath fallen shrubs and in dense vegetation or grass tussocks. Feeds on lizards, frogs and on small mammals, which is unusual in an Australian snake of this small size; most of our small elapids are lizard and frog eaters only. The Bardick is pugnacious when harassed, flattening the head, hissing and puffing with the forebody raised in an S-shape and striking repeatedly. Not generally regarded as dangerous, although a bite may cause severe local symptoms. Large individuals may be dangerously venomous, especially to small children. Bearing 3–14 live offspring in March or April, neonate SVL 82–125 mm, weight 1–2.5 g (Bush, 1992).

Black-naped snake

Neelaps bimaculatus (Duméril, Bibron & Duméril 1854)

Genus *Neelaps*
Günther 1863

This genus contains two species, both of which occur locally. They are small, very slender burrowing elapids, with broad head and neck bands but without multiple body bands. The anal and subcaudal scales are divided. Cogger (1992) and Storr *et al.* (1986) include all the members of this genus in *Simoselaps* and *Vermicella*, respectively.

Burns Beach

HABITAT

Occurs in sandy areas throughout the Perth region. Favours coastal and near-coastal dunes supporting low open vegetation and heath, as well as sandplain with banksia/eucalypt woodland. Sparsely distributed inland from the coast.

DESCRIPTION

A small, very slender snake with moderately depressed head and rounded snout. Dorsal scales smooth, shiny and in 15 rows at midbody. Ventrals 175–230, anal scale divided, subcaudals 15–25, all divided. Ground colour brown, red to bright-orange. Each scale has a cream blotch or base, giving the snake a reticulated appearance. Two broad black bands on head, one across nape and the other between the eyes and extending back from them. Snout and belly colour white to cream. Total length up to 45 cm.

GENERAL

Found in soft upper soil layer beneath leaf-litter and dense foliage of shrubs and tussocks. Also found inside abandoned ant nests and beneath logs and stumps. Displays a predilection for the decayed pulpy interior of old tree trunks. Feeds exclusively on skinks, in particular, small burrowing *Lerista* species. When it is harassed, the forebody is elevated with head acutely angled. It is believed to be an egg-layer, producing 2–6 in a clutch, but this has yet to be confirmed.

Black-striped snake
Neelaps calonotus (Duméril, Bibron & Duméril 1854)

City Beach

HABITAT — Inhabits coastal and near-coastal dunes, sandplain supporting heathlands and banksia/eucalypt woodlands. It is doubtful that it occurs on the Darling Range. A record of a sighting inland, from York (outside the Perth region), requires confirmation.

DESCRIPTION — A small, moderately slender snake having a snout that protrudes but lacks a cutting edge. Dorsal scales smooth, shiny and in 15 rows at midbody. Ventrals 125–143, anal scale divided, subcaudals 23–35, all divided. Ground colour is purple to bright orange-red, with each scale having a white to cream centre. A distinct feature in most individuals of this species is the bluish black vertebral stripe, which is up to three scales wide. The black scales along the back each retain the pale centre, forming a chain-like pattern extending from the neck to the end of tail. Most individuals have a continuous vertebral stripe, but it can be broken or (rarely) absent completely. There are usually remnants of stripe on the tail. A broad black band extends across the neck and another across the head, extending forward in line with the eyes. The tip of the snout is black, the interspace between this and the headband is white. Belly white. Total length 28 cm.
(Continued on next page)

Black-striped snake
(Continued)

Ellenbrook

GENERAL — This snake is one of Australia's smallest elapids. It feeds mainly on the Western Worm Lerista (*Lerista praepedita*) and the South-western Sandplain Worm Lizard (*Aprasia repens*), which are among the smallest burrowers found in its range. It is fascinating to watch it stalk its prey from beneath the sand. With an uncanny ability, it follows until within striking distance then, rising out of the sand, its head dives down and grasps its meal. No constriction has been observed, but the lizard seems to be immobilised rapidly by this little snake's venom. It has a very limited distribution and is found only on the Swan Coastal Plain between Mandurah and Lancelin. Because of the increasing urban and industrial development in its small range, we believe it should formally be recognised as threatened. It is an egg-layer, producing 2–5 eggs in a clutch (Shine, 1984). The smallest individual we have found came from Ellenbrook and measured SVL 101 mm, weight 0.72 g.

Western tiger snake

Notechis scutatus (Peters 1861) subsp. *occidentalis* Glauert 1948

Tiger snakes
Genus *Notechis*
Boulenger 1896

An endemic genus confined to southern Australia, including Tasmania and many offshore islands. It comprises two species and six subspecies, three of which are insular races found on islands. The dorsal scales are smooth, matt to moderately glossy and in 17–21 rows at midbody. The anal and subcaudal scales are undivided. The frontal scale is large and squarish, not, or hardly, longer than wide. The subocular scales are absent, internasals present. It is live-bearing (viviparous) with litters numbering as many as 64 (Fearn, 1993). An unconfirmed exceptional litter of 109 has been mentioned in herpetological circles for some years now.

Adult   Herdsman Lake

HABITAT — Common in vegetation bordering streams, swamps and lakes. Still occurs in the metropolitan area along the Swan and Canning Rivers. Large numbers are found on Carnac and Garden Islands.

DESCRIPTION — A large stout snake with broad blunt head. Dorsal scales smooth, matt and in 17 or 19 rows at midbody. Ventrals 140–165, undivided anal scale, subcaudals 36–51, all undivided. Ground colour black or dark-brown. The head may be dark olive-grey. Cream, yellow or orange bands, if present, may be narrow or broad and often indistinct. They are usually confined to the anterior half of body. Belly anteriorly cream or yellow with bold black markings, gradually changing to grey posteriorly. Total length 120 cm.

GENERAL — Diurnal and nocturnal. More tolerant of cold weather than other large elapids, often emerging on cool nights in search of food. Active during
(Continued on next page)

Western tiger snake
(Continued)

Juvenile Herdsman Lake

winter on fine sunny days, and occasionally on cool overcast days. Feeds principally on frogs, although lizards, mammals and young birds are also taken. Will ascend low vegetation in search of nesting birds. Like all snakes, it is an accomplished swimmer, and is often observed in water. Shelters beneath logs, tussock grass, low shrubs and in abandoned rabbit burrows. A Tiger Snake will attempt to flee if disturbed. They have a reputation for being aggressive, but this is a misinterpretation of their defensive attitude. It may occasionally stand its ground with forebody raised and neck flattened, but this defensive stance is nothing more than intimidation or bluff behaviour. It is not inclined to bite unless touched. Produces live young, litter sizes ranging from one to 40, neonates measure 15–24 cm and weigh 3.3–9.5 g. A captive female lived for 13 years and four months, and gave birth to 136 offspring in five consecutive litters, the first at 47 months of age.

DANGEROUSLY VENOMOUS. Venom contains components that cause both presynaptic and postsynaptic paralysis, coagulopathy and muscle destruction. An untreated bite can cause death. Juveniles should also be treated with caution.

Mulga snake
Pseudechis australis (Gray 1842)

Black snakes
Genus *Pseudechis*
Wagler 1830

A group of very large robust elapid snakes, having a broad depressed head. There are five species distributed throughout most of Australia, except in its southern extremities. One other species, *P. papuanus*, is endemic to New Guinea. The body scales are smooth, in 17–19 rows at midbody. Subcaudal scales vary considerably between individuals, but are usually undivided anteriorly and divided posteriorly. Combat has been recorded between male Mulga Snakes (*P. australis*).

Chittering

HABITAT — Uncommon locally, where it is confined to the northern Darling Range along the Avon River and to adjacent sandplains in the vicinity of Muchea and Yanchep. Sometimes encountered in the agricultural areas of that region. Recently recorded in the Bullsbrook and Red Hill areas. Perth is the southern limit of its range on the coast. It is widespread to the north and east of the Perth region, extending south to Narrogin.

DESCRIPTION — A large and moderately stout snake. Dorsal scales smooth, matt to moderately shiny and in 17 rows at midbody. Ventral scales 185–225, anal scale divided. Subcaudals 50–78, undivided anteriorly and divided posteriorly. Colour variable, from uniform pale-brown, olive, rusty or reddish brown to almost black. Darker specimens often have reticulated pattern from neck to tail, formed by a pale anterior portion of each scale. Belly cream to yellow, the forward edge of each ventral scale reddish brown. Recorded to 200 cm and occasionally reaching 300 cm in length.
(Continued on next page)

Mulga snake
(Continued)

Hatching of twins

GENERAL

Both diurnal and nocturnal. Feeds on lizards, snakes, including its own species, and mammals. Shelters in hollows, crevices, abandoned burrows, under rubbish such as corrugated iron or any other protected site. It is oviparous, laying 7–16 eggs in a clutch. These are 20–25 mm wide and 33–46 mm long, and weigh 11.5–14.5 g. They hatch after 65–88 days at 30–32°C. Neonate SVL 220–330 mm, weight 6–13 g. The smallest hatchlings recorded were SVL 192 mm and 197 mm, and weighed 4.18 g and 5.8 g, respectively. These two individuals hatched from a single egg. The combined weight of the two was equivalent to that of the single hatchlings in that clutch (Fyfe, 1991 and Bush, unpublished data). The Mulga Snake is often referred to as the King Brown Snake because of its large size and colour, but it is actually a member of the black snake genus. It seems to be extremely cold-tolerant. If provoked it may become very defensive, dorsally flattening the forebody in a striking stance. It has a tendency to hang on and chew when biting. It is literally Australia's most venomous snake, delivering more venom at a single bite than any other. Considered DANGEROUSLY VENOMOUS, although not involved in any fatalities between 1981 and 1994. The powerful venom contains no (or little) neurotoxin but contains strong myolysins, which cause severe muscle and tissue destruction in snakebite victims.

Dugite or spotted brown snake
Pseudonaja affinis Günther 1872

Brown snakes
Genus *Pseudonaja*
Günther 1858

This genus contains eight described taxa, which are found throughout most of Australia. These snakes are moderately small to large, relatively slender elapids. They have a narrow smallish head, not noticeably distinct from the neck, with large eyes, especially when immature. The dorsal scales are smooth, matt to very glossy and in 17–21 rows at midbody. The anal and subcaudal scales are divided.
(Continued on next page)

Mainland subspecies *affinis* — Canning Vale

HABITAT — Very common in the Perth region, favouring disturbed habitats. Has a predilection for human-made grasslands. A separate subspecies, *Pseudonaja affinis exilis*, occurs on Rottnest Island. It is smaller and consistently darker than Dugites on the mainland.

DESCRIPTION — A large slender to moderately stout snake with matt to slightly glossy dorsal scales in 19 rows at midbody. Ventrals 203–229, anal scale divided, subcaudals 50–65 (all divided). Ground colour highly variable, pale to dark-brown, yellow, reddish brown, olive, green or black. Often sparsely to heavily marked with irregular black spots or blotches. At one time the heavily spotted individuals were referred to by the Aboriginal name 'kabarda', while the term 'dugite' was applied to the uniformly coloured individuals. The head may be contrastingly paler or darker than the body. Occasionally individuals are found with distinct (rarely) or obscure body bands. Belly pale-grey, yellow or brown, spotted or blotched with darker grey, brown, orange or black. Juvenile colour olive-green, brown or yellow with the greater part of the head black. Belly whitish with small indistinct to distinct darker spots. Adult length to more than 200 cm.
(Continued on next page)

Dugite or spotted brown snake
(Continued)

The frontal scale has straight sides and is much longer than wide and about as wide as the supraoculars. Male brown snakes fight in the spring, but do little physical damage to each other. Brown snakes have small venom glands, but their venom is very toxic. It is known to cause coagulopathy, but the clinical symptoms are like those caused by an anti-coagulant, as only a portion of the coagulation cascade is affected. The victim may also become unconscious. Renal failure is common, and occasionally paralysis.

Rottnest Island subspecies *exilis* Storr 1989 — Rottnest Island

GENERAL An alert fast-moving diurnal snake that will forage at night during hot weather. This is the most common dangerous snake in the Perth area; 90 per cent of snakes found in backyards are Dugites. It is especially attracted to bird or rabbit enclosures and haysheds, where it hunts for mice. Population densities are relatively low in undisturbed bush remote from settlements, but are greater closer to developing areas, where there are more mice. Once an area is completely built-up the snakes have nowhere to live and the species becomes locally extinct. In addition to mice, it feeds on lizards, snakes and birds. Mating occurs in the spring and at this time males may be observed fighting. A female deposits one or sometimes two clutches numbering 10–30 eggs between November and February. They measure 25–45 mm long and 15–25 mm wide, and weigh 5–13 g. Hatching occurs from February to May, with neonate SVL 194–226 mm, weight 4.7–6.9 g (Bush, 1992). During late summer and early autumn large numbers of juveniles disperse from hatching sites. Many are found in the suburbs. These small snakes are best treated with some caution, as a bite can result in quite severe symptoms. Adult Dugites are usually quick to avoid people but will bite if trodden on, harassed, when cornered or handled. **DANGEROUSLY VENOMOUS.**

Adult Dugite — City Beach

Juvenile Dugite — Subiaco

Gwardar or western brown snake *Pseudonaja nuchalis* Günther 1858

Captive-bred snake, pale-headed form

HABITAT
Uncommon locally. It has been recorded on the Swan Coastal Plain near Midland and adjacent areas, preferring agriculturally disturbed land. In the Darling Range it has been recorded in the Susannah Brook area north-west of Gidgegannup. This is the most common snake around Toodyay, Northam and York. The area covered by this book represents the south-western limit of its range.

DESCRIPTION
A large slender agile snake. Dorsal scales smooth, moderately to very glossy in 17 rows at midbody. Ventrals 180–230, anal scale divided, subcaudals 42–65 (all divided). Adult length up to 160 cm. Coloration is highly variable, and several distinct colour forms occur with a multitude of intergrades:

Hooded form. Head and neck glossy black. Body grey, yellow, olive-green, pale-brown, orange-brown to rich reddish brown. Immature and young adults usually have each dorsal scale bordered with black along lower edge, forming a reticulated or herringbone pattern.

Captive-bred snake, banded form

Banded form. Head and neck may or may not be glossy black as in hooded form. If not, then pale to dark brown, with or without black scales on neck arranged in a 'V' or 'W' or randomly. Body with evenly spaced dark-brown or black bands, which are interspaced with silver-grey, pale-brown, yellow, orange or a distinctive red. Immature and young adults often have a herringbone pattern as well as three less distinct narrow bands on the pale interspaces. With age the herringbone pattern and the narrow bands fade. In many individuals the broad bands fade out anteriorly, leaving only the posterior ones. Bands do not extend fully onto the belly.

Pale-headed form. Head pale-brown, grey or bright-yellow. Body any shade of grey, olive or brown to almost black. If pale, then often a contrasting darker neck region or a few dark scales either random or more often arranged in a 'V'. Pale individuals often have a dark chevron or V-shape on neck, and immature and young adults have a dark herringbone pattern on body.
(Continued on next page)

Gwardar or western brown snake *(Continued)*

Hooded form — Ballidu

Uniform colour form. Similar to pale-headed form but head, neck and body are uniformly coloured with grey, fawn, olive, through any shade of brown to almost black. If black, then belly may be brilliant orange to almost reddish in striking contrast to black dorsum. Belly cream, yellow, orange to almost red with numerous darker spots or blotches. In spite of the vast colour variation among adults, juveniles (except when banded) are remarkably consistent in colour, with black head and neck. Sometimes the black is separated into two large blotches by a pale-brown area. Body light-brown to reddish brown with dark specks forming reticulated or herringbone pattern. Belly white to yellow, often with orange spots.

Uniform colour form   Carnarvon

GENERAL

Diurnal and nocturnal, depending on temperature. Inhabits most habitat types throughout its range. Around Midland it is generally confined to farming areas, where it utilises rubbish such as corrugated iron for shelter. It is attracted to these areas by the abundance of mice and rats. It also feeds on lizards and occasionally on other snakes. Like all members of the brown snake group, it is an egg-layer, producing up to 26 in a clutch. Healthy eggs are 25–44 mm long and 16–22 mm wide and weigh 4.8–10.7 g. If food is plentiful it may lay a second clutch 45–65 days after the first. Eggs hatch after 56–80 days at 30°C. Neonate SVL 150–235 mm, weight 2.3–7.7 g. A single clutch may include both banded and unbanded hatchlings (Bush, 1989a). These change considerably in colour as they grow (Bush, 1989b). If provoked, it will rear the forebody in an S-shape and defend itself. **DANGEROUSLY VENOMOUS** (see notes in introduction to genus).

Gould's hooded snake
Rhinoplocephalus gouldii (Gray 1841)

Wanneroo

Hooded snakes
Genus *Rhinoplocephalus* Müller 1885

A group of small, short-tailed elapid snakes with a broad depressed head. As most species have a black head, or remnants of this in the form of dark blotches, they are generally referred to as hooded snakes. Represented in Australia by 10 species, which are widely distributed throughout the continent, particularly in the south and east. One species extends north to New Guinea. Two species are found in the Perth region.
(Continued on next page)

HABITAT Common in the Perth region, especially the Darling Range. On the Swan Coastal Plain it favours areas supporting heathlands, blackboy (*Xanthorrhoea preissii*), eucalypt and banksia woodlands. In the Darling Range it is usually associated with granite and laterite outcrops.

DESCRIPTION Similar to the Black-backed Snake (*R. nigriceps*), but lacks the dark vertebral stripe. A small short-tailed snake with a distinct black head (rarely brown) and small black eyes. Body scales smooth, glossy and in 15 rows at midbody. Ventrals 153–177. Undivided anal scale. Subcaudals 25–38 (all undivided). Ground colour orange-brown, pinkish to reddish brown. Each scale finely edged with black, forming a reticulum. Head and neck glossy black with variably sized pale mark in front of each eye, sometimes these join to form a narrow transverse bar. Belly pearly white. Adult total length 55 cm.

The eyes are small and entirely dark. Dorsal scales smooth, glossy and in 15 or 17 rows at midbody. The anal and subcaudal scales are undivided. The frontal scale is straight-sided, slightly longer than it is wide and much wider than the supraocular scales. The nasal scale is entire and widely separated from the frontal scale.

Rhinoplocephalus gouldii and *R. nigriceps* were assigned to this genus by Storr *et al.* (1986), whereas Cogger (1992) places them in the genus *Suta*.

Ellenbrook

GENERAL An attractive, terrestrial and night-active snake that can occur in high densities on granite outcrops and areas where blackboys are abundant. Seeks concealment during the day beneath a range of cover including rocks, logs, fallen bark, human debris and inside stick-ant nests, termite mounds and dead blackboys. Nocturnal activity often brings it onto roads searching for skinks and geckos, which are its main food. Foraging snakes hunt for sleeping diurnal lizards. Produces live young in March or April, 3–7 in a litter. Neonate SVL 140–160 mm, weight 2.0–2.5 g. This snake has been observed on roads in the Stoneville area in June and July, suggesting that it is cold-tolerant. Often it is quite placid when first caught, although some individuals attempt to bite when handled. It is venomous but too small to be considered harmful. A bite is normally less painful than a bee sting, causing only mild local pain and swelling.

Black-backed snake
Rhinoplocephalus nigriceps (Günther 1863)

Karragullen

HABITAT — Common in jarrah forest in the Darling Range, particularly granite outcrops and areas dominated by blackboys. Uncommon on the coastal sandplain, where it is known from areas such as Mussel Pool and Gosnells.

DESCRIPTION — Similar to the Gould's Hooded Snake (*R. gouldii*), but has a dark vertebral stripe and darker less reddish body. A moderately robust member of the genus, with smooth glossy scales in 15 rows at midbody. Ventrals 108–200, anal scale undivided, subcaudals 18–40 (all undivided). Top of head and neck glossy black. A broad glossy dark-grey to black vertebral stripe or zone extends from neck to tip of tail. The flanks are orange-brown, each scale bordered by black, which together may form a reticulated pattern. Lips pale orange to white. Belly pearly white. Adult length to 60 cm.

GENERAL — A nocturnal snake that is tolerant of cool weather and therefore active throughout much of the year. Shelters under granite rocks, in decaying blackboys and termite mounds. Feeds on lizards and small snakes, including its own species. Produces up to five young in March or April. Neonate SVL 105–125 mm, weight 1.5–2 g (Bush, 1992). This species is usually pugnacious when handled, and may bite if given the opportunity. Venomous but not regarded as harmful. A bite from a mature specimen may cause mild pain and local swelling.

Banded sand snake or Jan's banded snake *Simoselaps bertholdi* (Jan 1859)

City Beach

Banded burrowing snakes
Genus *Simoselaps*
Jan 1859

A group of small, relatively robust short-tailed elapid snakes. None exceeds 40 cm in length and though venomous, they are considered harmless and do not usually attempt to bite. The combination of a depressed rounded or shovel-shaped snout, small eyes and smooth glossy scales make them efficient burrowers through leaf-litter and sand. Generally only encountered on the surface at night. The banded pattern is thought to deter or confuse predators as the snake moves in dim light.

HABITAT	Common on coastal dunes and sandplain supporting heath with banksia/eucalypt woodland. Seems to be absent from the Darling Range.
DESCRIPTION	A small robust attractive snake, with depressed head. Midbody scales in 15 rows, 112–130 ventrals, anal scale divided, subcaudals 15–25 (all divided). Ground colour yellow to orange with 18–31 broad black bands. The bands and interspaces are approximately equal in width. Pale interspaced scales usually bear darker edges, fading out on lower flanks. Head white to pale grey and densely flecked with dark brown, flecks merge on base of head to form a dark blotch. Belly white to pale yellow between the black bands. Total length up to 30 cm.
GENERAL	Mainly moves at night, although it has been seen active during the day (Browne-Cooper & Maryan, 1988). Shelters in loose topsoil beneath leaf-litter, rubbish, rocks and logs. Many of these harmless snakes are killed because they are mistaken for Tiger Snakes. When feeding, this species uses python-like constriction to restrain prey. It lays 1–8 eggs in a clutch (Shine, 1984). In late February, three newly hatched individuals were found in a spoil-heap. All were in a postnatal pre-slough condition, suggesting they had hatched within the past two weeks. Their average SVL was 92 mm, weight 1.16 g.

Narrow-banded snake
Simoselaps fasciolatus (Günther 1872) subsp. *fasciolatus*

Seabird

HABITAT Scarce in this region, probably the least common of the local burrowing snakes. On the Swan Coastal Plain favours sandy areas such as coastal and near-coastal dunes. Also found in banksia/eucalypt woodland over heath. Not known from the Darling Range.

DESCRIPTION A relatively robust snake, having a depressed snout with a sharp tip to assist burrowing. Midbody scales in 17 rows, ventrals 140–172, anal scale divided, 20–30 divided subcaudals. Ground colour white, cream to pale-grey, marked with numerous narrow irregular-edged brown to black bands. Bands never completely encircle body. Pale areas between bands bear reddish to blackish spots, which are either scattered or aligned in transverse series. A dark-brown to black blotch on the head is separated from a broader black blotch on the neck area by a narrow reddish brown band. Belly white. Total length to 40 cm.

GENERAL The shovel-shaped snout, where upper jaw protrudes over lower jaw, makes this snake an adept burrower. It can be found sheltering beneath leaf-litter on loose sand, in cracks in the ground and under decaying logs and stumps. Feeds primarily on lizards, particularly *Ctenotus* species, but has been recorded as taking reptile eggs (Shine, 1984). Believed to be an egg-layer, producing about five eggs in a clutch, but this has yet to be confirmed.

Southern half-girdled snake
Simoselaps semifasciatus (Günther 1863)

Ellenbrook

HABITAT Occurs throughout the Perth region, including the Darling Range. Found in most soils, from loose coastal sand to stony soils of the Darling Range.

DESCRIPTION A medium-sized robust snake with a distinct shovel-shaped snout bearing an acute cutting edge. Midbody scales in 17 rows, 147–188 ventrals. Anal scale divided, 14–26 divided subcaudals. Ground colour orange to reddish brown, fading towards lower flanks with numerous dark-brown to dark-grey bands. These never completely encircle the body and are narrower to broader than the pale interspaces. A broad dark-brown band present on head between, and extending back from, the eyes and a second band on the neck. Belly cream to white. Total length to 35 cm.

GENERAL Found in soil cracks, insect burrows and beneath logs, rocks and rubbish. Also found inside decaying logs, stumps and termite mounds. Shine (1984) found that it feeds exclusively on reptile eggs. No other Australian elapid has such a specialised diet. The dentition in this oophagous (egg-eating) snake is remarkably different from that of its close relatives. It has flat blade-like posterior maxillary teeth to penetrate egg shells. Feeding presumably occurs only in the warmer months in this region, when most reptiles are breeding. Believed to be an egg-layer, producing 2–5 eggs in a clutch.

SEA SNAKES

(Family Hydrophiidae)

SEA snakes mainly frequent the tropical and subtropical waters off our northern coast, but some are encountered near Perth. Although most of these are vagrants, the Yellow-bellied Sea Snake (*Pelamis platurus*) is very widespread and a true inhabitant of Perth waters.

Sea snakes are easily recognised. They have several external characteristics that are adaptations to marine life. They have laterally compressed bodies and paddle-shaped tails. They also have valvular nostrils, which close when the snake submerges. They can dive to considerable depths, and some species remain submerged for hours. Those individuals found on land have most often been trapped there by tidal changes and storms.

Most sea snakes are live-bearing, although a small number of species are egg-layers. Members of this latter group, the sea kraits (laticaudids), have broad belly scales and a less compressed body than other sea snakes. These characteristics allow excursions onto land for both basking and egg-laying. The young of viviparous species are born at sea. Clutch and litter sizes of sea snakes tend to be smaller than those of land snakes. Most sea snakes feed on fish, but at least one species feeds exclusively on fish eggs. Although they are generally placid creatures and their fangs are relatively small, they do possess powerful venom and should be treated as dangerously venomous. The purpose of venom in sea snakes, as with terrestrial snakes, is to subdue prey and aid digestion. They bite defensively only as a last resort when provoked.

Key to sea snakes

The key included here is only for those species that have been recorded as far south as the region covered by this book. The taxonomy used here follows Storr *et al.* (1986).

1	Ventral scales more than one-third of body width	**Shark Bay sea snake** (*Aipysurus pooleorum*)
	Ventral scales less than one-quarter of body width	2
2	Ventrals much smaller than dorsals, colour boldly black and yellow longitudinally	**Yellow-bellied sea snake** (*Pelamis platurus*)
	Ventrals almost as large as or larger than dorsals. If colour black and yellow, then transversely banded	3
3	Body stout and of fairly uniform thickness, ventrals usually fewer than 250	**Olive-headed sea snake** (*Hydrophis major*)
	Anterior part of body much more slender than posterior part, ventrals usually more than 250	4
4	Ventrals more than 350	**Bar-bellied sea snake** (*Hydrophis elegans*)
	Ventrals fewer than 340	5
5	Scale rows on neck more than 30, and on thickest part of body more than 43	**Spotted sea snake** (*Hydrophis ocellatus*)
	Scale rows on neck fewer than 30, and on thickest part of body fewer than 43	**Spectacled sea snake** (*Hydrophis kingii*)

Yellow-bellied sea snake
Pelamis platurus (Linnaeus 1766)

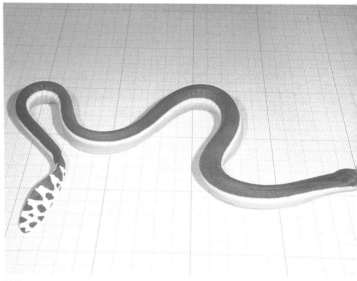

H. Ehmann   Cottesloe

HABITAT	Occasionally found on the beach and off the coast in the Perth region, especially in the Kwinana–Rockingham area. Also occurs further south.
DESCRIPTION	A moderately small slender sea snake with distinct coloration. Head long and narrow with large symmetrical scales. Midbody scales in 47–69 rows, 240–406 very small ventrals. Top half of body black or dark-brown, contrasting markedly with cream, yellow or pale-brown lower half. Tail pale yellow, spotted or barred with black. Total length up to 80 cm.
GENERAL	An ocean-inhabiting and wide-ranging sea snake. Sometimes observed floating on the ocean surface, where it will ambush fish. The species most commonly found washed-up on Perth beaches. **DANGEROUSLY VENOMOUS.**

Is it true that...?

THIS section has been compiled from questions that are commonly asked about reptiles.

Are any lizards venomous?

Yes. Although none of Australia's lizards are venomous, there are two in the world that are: the Gila Monster found in the Arizona Desert of North America, and its close relative the Mexican Beaded Lizard.

Which is the largest lizard in Australia, and which is the largest in the world?

Australia's largest lizard is the Perentie, which grows up to two metres long. The largest lizard in the world is the Komodo Dragon, found on Komodo Island and some surrounding islands of Indonesia. Both are monitor lizards.

Do Bobtails keep snakes away?
(Asked when this lizard is seen commonly in the garden.)

It is unlikely that Bobtails would keep snakes away, as the larger snakes feed on them. However, these lizards are of value because they feed on garden pests.

Do all lizards have ears?

No. Many of the smaller burrowers and some others do not have an ear-opening at all.

Which snake is the deadliest in the world?

The answer to this question varies, depending on how you measure 'deadliness'. If you use the total number of deaths from snakebite, the Asian and South American snakes must rank as the most deadly; the large venomous snakes such as the cobras, kraits and vipers of Asia and the lance-headed vipers of South America are responsible for thousands of deaths each year. But it is not possible to rank one snake above another in terms of relative deadliness, as there are too many variables to consider. Some factors that contribute to increased mortality from snakebite are: high human population densities in areas where snakes are common, tropical lifestyles, poor medical facilities and lack of antivenom. For example, in Sri Lanka there are 800 deaths from snakebite each year, whereas there are fewer than two in Australia. Sri Lanka has about the same number of people as Australia, but occupying an area 100 times smaller than that of Australia.

Which snake is the deadliest in Australia?

The brown snake group, which includes the Eastern Brown, Western Brown and Dugite, is thought to have been responsible for 15 of the 25 deaths from snakebite in Australia between 1981 and 1994. The tiger snake group ranks a close second, having been responsible for six deaths, followed by the Northern Taipan (two deaths) and the death adder (two deaths). The venom of the Inland Taipan has been shown in the laboratory to be very strong, but has never been known to have caused a death.

Do all venomous snakes have hollow fangs?

No, but all front-fanged snakes have fangs that are effectively hollow, similar to hypodermic needles. All sea snakes and all but five of our venomous land and freshwater snakes are front-fanged. The rear-fanged snakes, which are not common in Australia, have an open groove along the fang length.

How do I tell if a snake is venomous or nonvenomous?

If it is small and worm-like, it is nonvenomous. There are far more species of venomous than nonvenomous snake in Australia (115 of the 165 known species are venomous). As there is no consistent guide to separate the two groups, it is best to treat them all as being potentially venomous.

Do snakes shut their eyes?

No. The eye is protected by a transparent spectacle much like a fixed contact lens, which is renewed during the shedding of the skin.

Why do reptiles shed all their skin?

Reptiles are the only backboned animals that continue to grow throughout their lives. The outer skin is dead, therefore it will not grow with the reptile. So when the reptile gets too large for its skin, it is replaced. The skin is also subject to considerable wear. In burrowing lizards this wear is greatest on the nose, and they can shed this nose skin independently of the remaining skin.

Snakebite

Snakes and snakebite

Snakes are shy and therefore rarely seen. However, each spring and summer some do find their way into and around buildings and other areas where they come into contact with people. Remember though, they cannot eat you, therefore they don't want to bite you. The venom is a complex combination of proteins with two main functions: to immobilise prey and then to accelerate the digestion of that prey.

Snakes only bite in defence as a last resort, when threatened, accidentally picked up in material such as tree prunings and firewood, or trodden on. This is borne out by the fact that 95 per cent of bites are on the forearms or lower legs; most are on the lower legs.

Roughly 2000 people are bitten by snakes each year in Australia. Of these, 200 people require antivenom treatment and one or two bites prove fatal. This indicates the snake's reluctance to use venom for defence.

If first aid is carried out immediately after the bite, snakebites are rarely fatal. All actual or suspected snakebites should be treated as potentially dangerous and first aid administered.

There are 106 species of snake known from Western Australia, with 72 being front-fanged venomous snakes. Of these, 22 are

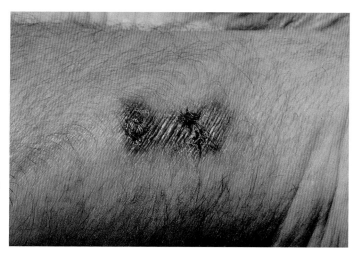

Snake-bitten arm showing tissue damage at bite site caused by Tiger Snake venom

sea snakes (hydrophids), with six species having been recorded as far south as the Perth region. Of the 50 venomous land snakes (elapids), 15 occur locally, but most are not dangerously venomous to adults. In the Perth area the Dugite or Spotted Brown Snake and the Western Tiger Snake are the most common elapids.

Australian elapid snakes are often said to be the deadliest or most venomous in the world, with the inference that this is to humans. The truth is, they are only the deadliest or most venomous if you are a mouse! In laboratory tests, mice are used as a standard to compare the strengths of different snake venoms (Broad *et al.*, 1979). It would be impossible to do these tests on humans, but we should bear in mind that the response in mice does not necessarily reflect what happens in humans. For example, tissue from mice and rats is 50 times less responsive to the venom of the Funnelweb Spider (*Atrax robustus*) than is human tissue (Underhill, 1988). Kellaway (1934) found adult mice to be unaffected by raw Funnelweb Spider venom. In fact, many small mammals are resistant to the venom of this spider, but humans can die from its effects.

Africa, Asia and South America have far fewer venomous snakes than Australia but have far higher rates of death from snakebite. India alone loses 5000–15 000 people a year. If Australia had a comparable number of fatalities relative to our population, we would have between 100 and 200 deaths a year. In fact, there are one or two deaths from snakebite a year in Australia. Between 1981 and 1994, only 25 deaths in Australia have been, in part, attributed to snakebite (Sutherland, 1992; media reports for 1993). Brown snakes (genus *Pseudonaja*) are believed to have been involved in 15 cases, tiger snakes (genus *Notechis*) in six cases, the Northern Taipan (*Oxyuranus scutellatus*) in two and death adder (genus *Acanthophis*) in two. All large venomous snakes are harmful, but there are too many variables involved to consider one species more of a threat than another.

Seasons and snakebite

Snakebite can happen at any time of the year, but snakes are less active during the cooler months (because a snake's body temperature is dependent on the external temperature). In the south-west of the State, there is a greater chance of accidental bites when the days start to warm up, from September to January. The cool nights at that time of year cause snakes to be lethargic, particularly in the morning, making them less responsive to approach. This is compounded by the winter proliferation of grasses; both humans and snakes are less visible when there is a lot of grass about.

Juvenile Dugites abound in February and March, and often find their way into buildings. They are typically greenish or brownish with dark head, and when disturbed they quickly adopt a defensive stance. Although unlikely to be dangerous to adults, a bite from one of these can cause quite severe swelling.

All elapids are venomous, but many are too small to be considered dangerous. Their bites cause no more discomfort than a bee or ant sting. About 75 per cent of Australia's elapids fall into this category. The venom of the larger species is more likely to be adapted for mammalian prey, and may therefore be more toxic to humans. Even the venom of juveniles of the larger species has this high toxicity. Immature individuals of the Gwardar or Western Brown Snake (*Pseudonaja nuchalis*) have been responsible for some of the most rapid deaths from snakebite in Australia. One such case occurred just north of Perth at Jurien. An adult male died in the car, trying to get to the local nursing post, shortly after being bitten. This was an exceptional case. If a victim of snakebite is treated within one hour, death is rare in Australia.

A quick reference guide to first aid for snakebite appears on pages 225–6.

First aid for snakebite

1. **Apply broad firm pressure bandage over bite and as much of limb as possible.** After covering bite continue bandaging up the limb and, if sufficient bandage, back down again.

2. **Immobilise limb with splint.** On leg, splint in straight position. On forearm, splint to elbow and support arm in sling.

3. **Keep victim still.** Bring transport to victim and take to nearest hospital.

   NOTE: **Do not wash venom from bite site!**

   For bites on trunk of body or face, apply local pressure only, with flat of hand.

Snakes and people

SNAKES are an integral part of the Australian bush. They are considered by many to be an unnecessary evil, but this is a misconception. As predators, snakes play an important part in the control of other animals. In many cases these other animals are introduced vermin such as rats and mice. Snakes pose little threat to sensible people. If sighted in a backyard and left alone, they will generally move on. If harassed, and if retreat is hindered by fences or walls, the snake may panic and feel threatened. Even at this stage the snake is reluctant to bite unless touched. Many bites occur when attempts are made to kill snakes.

Snakes are most likely to be found in backyards where old junk is stored directly on the ground, providing ideal shelter. For this reason, it is best to store junk off the ground. Bird aviaries and chicken yards attract mice, which in turn attract snakes.

Snakes that persist in backyard or workplace situations can be trapped with rat glue-traps. These are available from 'Robert Linton's' of Victoria Park.

There are few registered snake-catchers in the region. Some pest control companies will attend for a fee. The Department of Conservation and Land Management is reluctant to send wildlife officers to remove snakes, but will provide the telephone numbers of registered snake-catchers.

Dogs, cats and snakes

DOGS come into contact with venomous snakes because of their behaviour. In particular, puppies and young dogs are inquisitive and always looking for fun. The snake may be no more than a rustle in the grass, but the dog cannot resist chasing the sound and pouncing up and down in the grass in search of the source. If the snake is out in the open, the dog might consider it something to play with. In either case the snake might bite in self-defence.

Older dogs are similarly at risk, but they are generally less interested in playing. Older dogs are more likely to exhibit hunting or territorial behaviours, although these are to some extent breed-specific. If a snake moves through the yard these dogs are quick to attack it. In many cases they are successful, but as a dog get older its reflexes slow down, and there is an increasing likelihood of snakebite.

Cats are natural hunters and often find snakes. A problem to owners of snake-catching cats is their habit of returning to the house with a live snake. Little can be done to discourage cats from hunting snakes.

Precautions

Enclosures or backyards where dogs are kept in outer suburban and rural areas should be free of long grass and rubbish. This also applies to the area surrounding those enclosures. The addition of a low (up to one metre) continuous corrugated iron or cement sheet wall around the perimeter of the enclosure will generally keep snakes out. Dry dog food should be kept carefully, so as to avoid attracting mice.

When exercising a dog in bushland during spring, summer and autumn it is advisable to restrain it on a leash. If the dog must be run, choose an open area where visual contact can be maintained at all times.

Dogs are more susceptible to snake venom than are cats. If a dog is bitten by a large venomous snake, then the prognosis is rarely good. But if there are signs of life, with the correct antivenom, a dog can make a remarkable recovery. When working farm dogs are bitten, they often die in a vehicle on the way to town for treatment.

Initial symptoms of snakebite in dogs and cats

Your local veterinary surgeon should be aware of the types of snakes found in the district, and would generally have access to the antivenoms for these. In humans, a swab of the bite-site will allow identification of the specific type of antivenom. But with dogs (and other hairy animals) it is often difficult to locate the bite-site, therefore, some appreciation of the symptoms can help in identifying the antivenom needed.

In the Perth area, two common snakes are often involved with bites to dogs and cats. One is the Dugite or Spotted Brown

Snake (*Pseudonaja affinis*), a very mobile nomadic snake that has small fangs and produces small amounts of very toxic venom. Often the symptoms are slow in onset and include a progressive paralysis commencing at the back legs and extending forward. After a bite from a Dugite, it is not uncommon to observe a dog or cat having problems supporting the hind part of the body. Eventually paralysis becomes total, with a lolling of the tongue, non-responsive pupils and laboured breathing. Brown Snake antivenom is used for bites from the Dugite and the Gwardar or Western Brown Snake (*Pseudonaja nuchalis*).

Western Tiger Snakes (*Notechis scutatus*) are also often involved with bites to dogs and cats. Dogs bitten by this species (and perhaps by any species other than a brown snake) become very agitated immediately after the bite and display hyper-activity for a brief period, before collapsing in an unconscious state with lolling tongue and laboured breathing. Although cats may display a similar response, they are more likely to seek out a quiet place and rest. The animal becomes progressively more lethargic until death or recovery.

Although Common Death Adders (*Acanthophis antarcticus*) and Mulga Snakes (*Pseudechis australis*) are widespread in Australia, they are not common locally and are therefore less likely to be encountered.

If the actual bite is observed and is on a limb, apply a pressure bandage to retard absorption of venom. For bites on the body apply local pressure with the hand. If the snake has been found and killed, take it with the dog for identification. If the dog is in an advanced state of collapse do not hesitate in getting it to a veterinary surgeon.

Bibliography

Here is a list of books on Australian reptiles and frogs. Most should be available from the larger libraries and bookshops.

Bush, B. (1981). *Reptiles of the Kalgoorlie-Esperance Region.* (B. Bush, Perth.)
Cann, J. (1978). *Tortoises of Australia.* (Angus & Robertson, Sydney.)
Cogger, H.G. (1992). *Reptiles and Amphibians of Australia.* (5th Edition) (A.H. & A.W. Reed, Sydney.)
Gow, G. (1989). *Complete Guide to Australian Snakes.* (Angus & Robertson, Sydney.)
Greer, A.E. (1989). *The Biology and Evolution of Australian Lizards.* (Surrey Beatty & Sons, Sydney.)
Heatwole, H.F. (1987). *Sea Snakes.* (New South Wales University Press, Sydney.)
Kwinana, Rockingham & Mandurah Branch of the WA Naturalists' Club (1988). *A Guide to the Flora and Fauna of the Rockingham Offshore Islands and Cape Peron.* (WA Naturalists' Club, Perth.)
Shine, R. (1991). *Australian Snakes—A Natural History.* (Reed, Sydney.)
Storr, G.M., Smith, L.A. & Johnstone, R.E. (1981). *Lizards of Western Australia I: Skinks.* (WA Museum, Perth.)
Storr, G.M., Smith, L.A. & Johnstone, R.E. (1983). *Lizards of Western Australia II: Dragons and Monitors.* (WA Museum, Perth.)
Storr, G.M., Smith, L.A. & Johnstone, R.E. (1986). *Snakes of Western Australia.* (WA Museum, Perth.)
Storr, G.M., Smith, L.A. & Johnstone, R.E. (1990). *Lizards of Western Australia III: Geckos and Pygopods.* (WA Museum, Perth.)
Sutherland, S.K. (1983). *Australian Animal Toxins.* (Oxford University Press, Oxford.)
Tyler, M.J., Smith, L.A. & Johnstone, R.E. (1984). *Frogs of Western Australia.* (WA Museum, Perth.)
Tyler, M.J. (1989). *Australian Frogs.* (Penguin Books, Ringwood.)
Wilson, S.K. & Knowles, D.G. (1988). *Australia's Reptiles.* (Collins, Sydney.)

There are a number of more specialised articles and reports on the reptiles and amphibians of the Perth area. These are rarely stand-alone publications, but are included in scientific journals or in the published results of more comprehensive surveys.

How, R. & Dell, J. (1990). Vertebrate fauna of Bold Park, Perth. *West. Aust. Nat.* **18** (4/5): 122–131.

Maryan, B. (1984). *Delma grayii* in an urban area near Perth, Western Australia. *Herpetofauna* **16** (1): 29.

Maryan, B. (1993). Herpetofauna of an urban area near Perth, Western Australia. *West. Aust. Nat.* **19** (2): 174–183.

Robinson, D., Maryan, B. & Browne-Cooper, R. (1987). Herpetofauna of Garden Island. *West. Aust. Nat.* **17** (1): 11–13.

Turpin, M. (1990). Ecological appraisal of an isolated banksia woodland Reserve No. 3694 south of the Swan River, Perth. *West. Aust. Nat.* **18** (4/5): 131–138.

Wellington, B. & Dell, J. (1989). Reptiles and amphibians of the Darling Scarp near Perth. *West. Aust. Nat.* **17** (8): 226–228.

References

The following articles have all been referred to elsewhere in this book in abbreviation (the author's name followed by the date the article was published). Here, we have given enough information to allow you to seek out a copy of the reference. You will find that the author's name is followed by the date of publication, title, periodical in which the article was published (in italics), volume number (in bold) and page number.

Bamford, M.J. (1992). Growth and sexual dimorphism in size and tail length in *Tympanocryptis adelaidensis adelaidensis* (Lacertilia: Agamidae). *Herpetofauna* **22** (2): 25–30.
Broad, A.J., Sutherland, S.K. & Coulter, A.R. (1979). The lethality in mice of dangerous Australian and other snake venom. *Toxicon* **17**: 661–664.
Browne-Cooper, R. (1992). A record of aggregation in *Lerista elegans* (Lacertilia: Scincidae). *Herpetofauna* **22** (2): 38–39.
Browne-Cooper, R. & Maryan, B. (1992). Notes on the status of the skink lizard, *Ctenotus lancelini* on Lancelin Island. *West. Aust. Nat.* **19** (1): 63–64.
Browne-Cooper, R. & Maryan, B. (1988). Record of diurnal feeding in *Vermicella bertholdi* (Jan, 1859) (Serpentes: Elapidae). *Herpetofauna* **18** (2): 25.
Bush, B. (1983a). A record of reproduction in captive *Delma australis* and *D. fraseri* (Lacertilia: Pygopodidae). *Herpetofauna* **15** (1): 11.
Bush, B. (1983b). Notes on reproduction in captive *Menetia greyii* (Lacertilia: Scincidae). *West. Aust. Nat.* **15** (6): 130.
Bush, B. (1984a). Male–male combat in the Western Banjo Frog *Lymnodynastes dorsalis* (Gray). *Herpetofauna* **15** (2): 43.
Bush, B. (1984b). Seasonal aggregation behaviour in a mixed population of legless lizards, *Delma australis* & *D.fraseri*. *Herpetofauna* **16** (1): 1.
Bush, B. (1988). An unsuccessful breeding record for the Western Carpet Python, *Morelia spilota imbricata*. *Herpetofauna* **18** (1): 30.
Bush, B. (1989a). Polymorphism in captive-bred siblings of the snake, *Pseudonaja nuchalis*. *Herpetofauna* **19** (20): 28.
Bush, B. (1989b). Ontogenetic colour change in the Gwardar, *Pseudonaja nuchalis*. *West. Aust. Nat.* **18** (2): 25.
Bush, B. (1992). Some records of reproduction in captive lizards and snakes. *Herpetofauna* **22** (1): 26.
Daly, G. (1992). Reproductive biology of the Scalyfoot *Pygopus lepidopodus*. *Herpetofauna* **22** (2): 40–41.
Dell, J. (1985). Arboreal Geckos feeding on plant sap. *West. Aust. Nat.* **16**(4): 69.

Fearn, S. (1993). The Tiger Snake *Notechis scutatus* (Serpentes: Elapidae) in Tasmania. *Herpetofauna* **23** (2): 17–29.

Fyfe, G. (1991). Captive breeding of Mulga Snakes (*Pseudechis australis*) from Central Australia. *Herpetofauna* **21** (2): 36.

Hutchinson, M.N., Donnellan, S.C., Baverstock, P.R., Krieg, M., Simms, S. & Burgin, S. (1990). Immunological relationships and generic revision of the Australian lizards assigned to the genus *Leiolopisma* (Scincidae: Lygosominae). *Aust. J. Zool.* **38**: 535–554.

Kellaway, C.H. (1934). A note on the venom of the Sydney Funnelweb Spider, *Atrax robustus. Med. J. Aust.* **1**: 678–679.

King, D. & Green, B. (1979). Notes on diet and reproduction of the Sand Goanna, *Varanus gouldii rosenbergi. Copeia* (**1**): 64.

Kluge, A.G. (1993). *Aspidites* and the phylogeny of Pythonine snakes. *Rec. Aust. Mus.* Supplement **19**: 1–77.

Lee, A.K. (1967). Studies in Australian amphibia II. Taxonomy, ecology and evolution of the genus *Heleioporus* Gray (Anura: Leptodactylidae). *Aust. J. Zool.* **15**: 367–439.

Main, A.R. (1957). Studies on Australian amphibia I. The genus *Crinia* Tschudi in south Western Australia and some species from south-eastern Australia. *Aust. J. Zool.* **5**: 30–55.

Main, A.R. (1965). *Frogs of Southern Western Australia.* (WA Naturalists' Club, Perth.)

Maryan, B. (1987). Notes on reproduction in captive *Lialis burtonis. West. Aust. Nat.* **16** (8): 190.

Maryan, B. (1988). Notes on reproduction in captive *Ramphotyphlops australis* (Gray). *Herpetofauna* **18** (2): 1.

Roberts, G.A. (1981). Terrestrial breeding in the Australian leptodactylid frog *Myobatrachus gouldii* (Gray). *Aust. Wildl. Res.* **8**: 451–462.

Shea, G.M. (1991). Revisionary notes on the genus *Delma* (Squamata: Pygopodidae) in South Australia and the Northern Territory. *Rec. S. Aust. Mus.* **25** (1): 71–90.

Shea, G.M. & Peterson, M. (1993). Notes on the biology of the genus *Pletholax* Cope (Squamata: Pygopodidae). *Rec.West. Aust. Mus.* **16** (3): 419–425.

Shine, R. (1984). Ecology of small fossorial Australian snakes of the genera *Neelaps* and *Simoselaps* (Serpentes, Elapidae). pp 173–183 in: *Vertebrate Ecology and Systematics—A Tribute to Henry S. Fitch.* Edited by R.A. Seigel, L.E. Hunt, J.L. Knight, L. Malarel & N.L. Zuschlag. (Special publ., University of Kansas.)

Smith, L.A. (1985). A revision of the *Liasis childreni* species-group (Serpentes: Boidae). *Rec. West. Aust. Mus.* **12** (3): 257–276.

Sutherland, S.K. (1992). Deaths from snake bite in Australia, 1981–1991. *Med. J. Aust.* **157**: 740–745.

Underhill, D. (1988). *Australia's Dangerous Creatures.* p 170. (Readers Digest, Sydney.)

Wells, R.W. & Wellington, C.R. 1983. A synopsis of the class Reptilia in Australia. *Aust. J. Herp.* **1** (3 & 4): 73–129.

Worrell, E. 1963. *Reptiles of Australia.* (Angus & Robertson, Sydney.)

Glossary

aestivation: dormancy during summer or dry season.
Agamidae: dragon lizard family.
aggregate: a type of behaviour, where a species congregates at a site.
agile: quick and light in movement.
Amphibia: class comprising all amphibians.
amplexus: the sexual embrace of amphibians.
anal scale: one or more scales covering the anterior margin of the vent in snakes.
annulus: narrow ring-like structure.
anterior: refers to front end of body or tail.
anticoagulant: component of venom that impedes blood-clotting.
antivenom: serum prepared from animal blood, containing natural proteins that neutralise venom. Once known as 'antivenene', but this term is not recognised today.
apical: pertaining to the tip.
apical plates: adhesive scales on the underside and at the tips of digits in geckos.
aquatic: inhabiting water.
arboreal: inhabiting trees.
atypical: not typical, abnormal.
axilla: armpit.
azygous: not paired.
band: a transverse marking, longer than a bar.
bar: a short transverse or longitudinal marking.
base colour: see **ground colour**.
bicarinate: bearing two keels.
bifid: divided by a deep cleft. Usually refers to the 'forked' condition of a snake's tongue.
blotch: an irregularly shaped marking, larger than a spot.
Boidae: python and boa family.
callose: refers to scales on underside of digits when bearing calli.
callus: (plural **calli**) a thickening of a subdigital scale to form a raised ridge wider than a keel.
cannibalistic: feeding on members of the same species (or family).
canthus rostralis: longitudinal ridge located between top and side of snout.
carapace: top shell of turtles and tortoises.
carinate: keeled (see **keel**).

caudal: pertaining to the tail.
Chelidae: freshwater turtle family.
Cheloniidae: sea turtle family.
chevron: a V-shaped mark.
circumorbital: see **circumocular**.
circumocular: around the eye.
class: a taxonomic group containing one or more orders.
cloaca: common opening for reproduction and excretion, in amphibians, reptiles and birds.
cloacal spurs: a pair of tubercles, spines or clusters of spines located on each side of the cloaca or vent.
cloacal tubercles: see **cloacal spurs**.
clutch: the eggs laid as a group at any one time.
coagulant: component of venom that triggers off part or all of the blood-clotting process.
coagulation cascade: the whole complex blood-clotting process.
coagulopathy: an abnormal effect on the natural blood-clotting process.
coalesce: align.
Colubridae: family of solid-toothed and rear-fanged snakes.
community: a group of plants and animals living in the same environment.
compressed: flattened laterally or from the side.
conical: cone-shaped (in relation to scales).
copulation: sexual union.
crepuscular: active during sunrise or sunset.
crest: a longitudinal row of raised scales along midline of neck, back or tail.
cryptic: coloration that enables an animal to remain hidden or camouflaged.
cryptozoic: habit of living in concealed or hidden places, e.g. beneath rocks or leaf-litter.
cytotoxin: component of venom that causes destruction of blood and tissue cells.
dash: a short linear marking.
depressed: dorsally flattened body shape.
diffuse: pattern broken or obscure, merging with ground colour.
digits: fingers and toes.
dilated: the expansion of digits at the end or base.
dimorphism: having two distinguishable forms, see **sexual dimorphism**.
Diplodactylinae: subfamily within the Gekkonidae family.
disc: round terminal pad at end of digits.
distal: furthermost extremity from point of attachment.
diurnal: active by day.
dorsal: pertaining to the back or upper surface.
dorsolateral: immediately beneath the boundary between back and side.

dorsoventral: immediately above the boundary between upper and lower surface.
dorsum: back or upper surface.
duct: tube.
ecology: the branch of biology that deals with the interrelations between organisms and their environment.
ectothermic: body temperature regulated by behaviour, e.g. moving in and out of sun; 'cold-blooded'.
elapid: see **Elapidae**.
Elapidae: fixed front-fanged venomous land snake family.
elliptic: oval-shaped.
endangered: threatened with extinction unless given special protection.
endemic: confined to a specific region.
endothermic: body temperature regulated internally; 'warm-blooded'.
entire: not divided (in reference to the anal and subcaudal scales in snakes).
environment: the external surroundings in which a plant or animal lives.
ephemeral: seasonal (in reference to wetlands that are not full of water throughout the year).
excavate: to dig.
fangs: enlarged teeth each side of upper jaw in venomous snakes. If situated towards the back of mouth, then venom canal is an open groove as in some members of the Colubridae. If at the front of the mouth, venom canal is enclosed, as in the Elapidae and Hydrophiidae.
fauna: animals indigenous to a region.
femoral: pertaining to the thighs.
femoral pore: opening of a duct in or between scales on the thigh.
flanks: sides of body.
flaps: a pair of remnant hindlimbs in legless lizards, located on either side of vent.
fossorial: living or burrowing in topsoil or leaf-litter.
fragile: the condition of tail in some lizards, where tail can be broken off but will regrow.
fragmented: non-symmetrical or irregular (in reference to scalation).
frontal: large symmetrical head scale located on midline between the eyes.
frontonasal: scale or scales on snout, located behind internasals in many lizards and turtles.
frontoparietals: scale or scales located between frontal and parietals in lizards.
fused: joined (with reference to eyelids, scales or markings).
Gekkonidae: gecko family.
genera: plural for genus.
generic: pertaining to genus.
genus: formal group of closely related species.

glossy: shiny.
granular: small uniform scales.
ground colour: background colour, as distinct from pattern or markings.
gular fold: transverse fold of skin on throat of some agamids.
gulars: scales covering throat.
habitat: place within an environment where an organism lives.
haemotoxin: component of venom that destroys red blood cells.
hatchling: juvenile animal newly hatched from egg.
heath: low scrubby vegetation over sandy soil.
hemipenes: pair of independent male sex organs in lizards and snakes.
herpetofauna: reptile and amphibian fauna.
herpetology: the study of reptiles and amphibians.
heterogeneous: varied in size and shape.
homogeneous: uniform in size and shape.
Hydrophiidae: family of sea snakes commonly called 'true sea snakes' because of their limited ability or inability to move on land. All are live-bearers.
imbricate: overlapping (in relation to scales).
insular: refers to an isolated form or race, especially on an offshore island.
internasals: scales located between nasals immediately behind rostral.
interorbitals: scales located on head between eyes.
interparietals: scales located on head between parietals.
irregular: random, not aligned.
invertebrate: an animal without a backbone.
Jacobson's organ: a sensory structure located in roof of mouth in many reptiles, onto which the tongue is touched after sampling the surrounding air.
juxtaposed: side by side, but not overlapping.
keel: a narrow longitudinal ridge, usually on a scale.
labials: scales on lips, excluding tip of snout.
Lacertilia: suborder comprising all lizards.
lamellae: scales on underside of digits.
lanceolate: angular or pointed (in relation to scales).
larva: (plural **larvae**) an immature form that develops into a different adult form by metamorphosis.
lateral: the side.
laterodorsal: outermost part of back, immediately before side.
Laticaudidae: family of banded sea snakes commonly called 'sea kraits'. Members spend considerable time on land and are egg-layers.
lobules: small lobes protruding across ear-opening. Usually on front edge only.
longitudinal: aligned length-wise along the body.

lore: region between nostril and eye.
loreals: scale or scales located between nasal and preoculars.
loreotemporal stripe: stripe extending from lore through eye to temple.
mandible: lower jaw.
marine: inhabiting the sea.
matt: lacking shine.
maxilla: upper jaw.
maxillary teeth: teeth of upper jaw.
median: middle or midline.
mental: scales located at tip of lower jaw.
metamorphosis: the transformation of an animal from larva to adult, e.g. tadpole to frog.
metatarsal tubercle: raised structure on inner side of foot in frogs.
microhabitat: specific place occupied by an organism within its habitat.
midlateral: midline of flank.
midbody scales: located at middle of body and counted around body.
monogeneric: a family composed of a single genus.
monotonal: uniform colour, without markings or pattern.
monotypic: a genus comprised of a single species.
morphology: body shape, external structure or characteristics.
mottled: heavily spotted or blotched.
mucron: a sharp point (shorter than a spine) at distal end of some keels.
mucronate: bearing a mucron.
multicarinate: bearing more than three keels.
Myobatrachidae: family comprising the 'southern frogs'.
myolysin: antibody in venom that destroys muscle tissue.
nape: back of neck.
nasal: scale in which the nostril is located.
nasal cleft: groove extending through or terminating at nostril in blind snakes.
neurotoxic: toxic to the nervous system.
neonate: newly born offspring.
niche: the status of an animal within its community, which determines its activities and relationships with other organisms.
nocturnal: active by night.
non-fragile: tail not readily broken.
nuchal: back of head or neck; also scales in this region that are wider than adjacent dorsals and usually arranged in two series.
nuptial spine: spike on fingers of males in some frogs.
oblique: not straight.
occipitals: scales on head behind parietals in some lizards, e.g. genus *Tiliqua*.

ocelli: (singular **ocellus**) small pale eye-like markings with dark edging.
omnivorous: feeding on plants and animals, both dead and living.
ontogenesis: the growth and development of an animal.
oophagous: feeding on eggs.
orbit: eye socket.
order: taxonomic subdivision of class.
oviparous: egg-laying.
ovum: (plural **ova**) an unfertilised female reproductive cell.
palmar: sole of front foot.
palpebral disc: transparent area in lower eyelid of some skinks.
palpebrals: small scales on eyelids.
paratoid gland: (also **parotoid**) a swollen glandular region on the head of some frogs and toads. It usually commences just behind the eye, often extending onto the neck.
paravertebral: a longitudinal region along back, parallel to vertebral zone.
parietal eye: a small spot-like light-receptive organ on midline of parietal region in some lizards.
parietals: large head scales behind frontal.
parthenogenic: refers to an all-female species. Eggs of such species develop into new individuals without fertilisation from a male.
pelagic: ocean-inhabiting.
plantar: sole of rear foot.
plastron: bottom shell of turtles and tortoises.
pleistocene: geological era lasting from about 1.5 million years ago until 10 000 years ago.
pleurodirous: a group of turtles that withdraw the head beneath the carapace by horizontal bending of the neck.
poikilothermic: see **ectothermic**.
postanal: behind vent or cloaca.
posterior: refers to back end of body or tail.
postmental: median scale located behind mental scale in certain geckos.
postnasal: small scale located just behind nasal.
postnatal: period shortly after birth or hatching.
postocular: small scales immediately behind eye.
postsynaptic: refers to snake venom neurotoxins that act on muscle cell membrane and block signals from the brain, causing paralysis.
preanal: immediately forward of the vent or cloaca.
preanal pores: pores located immediately forward of cloaca.
prefrontal: scales on head immediately forward of frontal scale.
prehensile: refers to a tail that can grasp to aid climbing.
prenasal: small scales located immediately forward of nostril.

preocular: scales immediately forward of eye.
preslough: condition in reptiles just prior to shedding of the skin. Often associated with temporary dulling of coloration and clouding of cuticle scales.
presuboculars: scales on side of head located just behind the loreals.
presynaptic: refers to snake venom neurotoxins that act on nerve cell membrane and block signals from brain, causing paralysis.
proximal: towards the point of attachment of an appendage or structure.
pugnacious: quickly defensive.
pupil: dark aperture at centre of eye. May be round, or a vertical or horizontal slit.
Pygopodidae: family comprising all Australian legless lizards.
pygopods: (also **pygopodids**) legless lizards.
quadrilateral: large shield-like scale on back of head, formed by fusion of parietal scales.
race: see **subspecies**.
reticulated: see **reticulum**.
reticulum: a net-like pattern.
retractile: able to be retracted (with reference to claws).
rhomboidal: diamond-shaped.
robust: stout bodied.
rostral: scale covering tip of snout on upper jaw.
rugose: rough or wrinkled.
Salientia: order comprising all frogs.
scalation: arrangement, size and shape of scales.
scapular: referring to the shoulder.
Scincidae: family comprising all skink lizards.
Serpentes: suborder comprising all snakes.
serrate: saw-like in profile.
sex ratio: relative proportion of males to females in a population.
sexual dimorphism: morphological or colour differences between males and females of a species within a population.
shield: enlarged scale on the head.
slough: the voluntarily discarded skin.
snout–vent length: (abbreviated **SVL**) length from tip of snout to centre of vent.
sp.: (plural **spp.**) abbreviation for **species**.
spawn: frog eggs.
species: a formal group of related organisms, which freely interbreed in their natural state to produce sexually viable offspring. A species is referred to by its scientific name consisting of two words: the genus name followed by the species name, e.g. *Aprasia pulchella*.
speck: small dot of colour.

spectacle: a transparent scale covering the eye.
spine: scale with raised, sharp point.
spinose: bearing a spine or spines.
Squamata: order containing all snakes and lizards.
stripe: broad longitudinal line.
striate: refers to scales bearing longitudinal grooves.
subcaudals: scales on the underside of tail.
subdigital: pertaining to lower surface of fingers or toes.
subfamily: formal subdivision of family.
subnasals: scales located between nasal and upper labial.
suborder: subdivision within an order e.g. suborder Serpentes for snakes.
subspecies: a subdivision within a species. Usually evolved because of geographic isolation. Also termed 'geographic race'.
substrate: ground on which an animal lives.
suffuse: refers to pattern becoming obscure.
supracaudals: scales located on top of tail.
supraciliaries: scales located above eye.
supralabial: immediately above lips.
supraloreal: large scale located between anterior frontal and loreals in certain pygopods.
supranasal: scale located above nasal in some skinks, e.g. most *Morethia* spp.
supraoculars: scales on head directly above eye.
sutures: grooves between non-overlapping scales.
SVL: see **snout–vent length**.
sympatric: two or more species occurring together in the same region.
tail length: length measured from centre of vent to tip of tail. A regenerated tail is not measured because it is shorter than the original.
taper: becoming narrow or pointed.
taxon: (plural **taxa**) any formal unit of classification of organisms, e.g. species or family.
taxonomy: the science of classification of plants and animals.
temple: area on side of head between ear and eye.
temporals: large scales located on temples.
terrestrial: living on the ground.
Testudines: order comprising turtles and tortoises.
thermoregulation: ectothermic animals' control of their body temperature (usually by means of behaviour).
torpor: period of winter inactivity of reptiles.
total length: the sum of **SVL** and **tail length**.
transparent disc: see palpebral disc.
transverse: across the body or tail.

trihedral: triangular.
trilobed: refers to the condition in some blind snakes (genus *Ramphotyphlops*) where snout has three lobes when viewed from above.
tubercle: small circular protuberance or spine on the skin or scales.
tubercular: bearing tubercles.
tympanum: eardrum.
type locality: location where a species or subspecies was originally collected, and description on the basis of that specimen.
Typhlopidae: family of worm-like burrowing snakes known as blind snakes.
upper labials: scales covering the upper lips, excluding scale at tip of snout.
upper lateral: longitudinal region or line on upper part of flanks.
varanidae: the goanna or monitor family.
variegated: a pattern composed of a variety of colours and markings.
venom: modified saliva of some snakes, principally used for subduing and digesting prey. A specialised enzyme.
venomous: refers to snakes possessing venom glands, ducts and fangs.
vent: transverse opening of the cloaca.
venter: see **ventral**.
ventral: refers to belly or scales of belly.
ventrolateral: longitudinal region or stripe located on lower side of body adjacent to lower surface.
vertebral: longitudinal line or region along midline of back.
vertebrate: any animal with a backbone.
viviparous: live-bearing.
vomerine teeth: tooth-like structures found in the upper jaw (in reference to frogs).

Index

Bold numbers represent pages for descriptions and/or photos.

A
***Acanthophis* 164**, 200
  Acanthophis antarcticus 12, **164, 165**, 206
Aclys 55, **82**
  Aclys concinna **82**
adelaidensis 11, 14, 24, **44, 97, 98**
affinis 15, **179-181**, 206
Aipysurus pooleorum 193
alboguttatus **67**
albopunctatus **34**
amethistina 156
Amethyst Python 156
antarcticus 12, **164, 165**, 206
Antaresia **159**
  Antaresia childreni 159
  Antaresia perthensis 156
  Antaresia stimsoni **159**
Aprasia 55, 79, 80, **83**
  Aprasia repens 11, 83, **84**, 174
  Aprasia pulchella 13, **83**, 84
Arenophryne rotunda 22
Aspidites 156, 157
  Aspidites ramsayi **20**
australis 12, **152, 153, 177, 178**, 206

B
Banded Burrowing Snake **189**
Banded Sand Snake **189**
banded skinks **127**
Bar-bellied Sea Snake 193
Bardick 10, **170, 171**
Barking Gecko **78**
barroni **156**
barycragus **35**
Bassiana **110**
  Bassiana trilineata 14, **110, 111**
Beaked Blind Snake **155**
bertholdi **189**

bimaculatus **172**
binoei 62, **72**
Binoe's Prickly Gecko 62, **72**
Black-backed Snake 13, 186, **188**
Black-naped Snake **172**
Black-necked Cobra 162
black snakes **177**
Black-striped Snake 11, 19, **173, 174**
Black-tailed Monitor **104, 105**
blind snake 8, 12, **149**
Bluetongue 10
  Western **140**, 142
  Western Slender **121**
Bobtail 3, 15, 16, **141, 142**, 195
Bold-striped Four-toed Lerista **131**
branchialis 10, **121**
brevicauda 99
Broad-banded Sandswimmer **127**
brown snakes **179**
Bufo marinus 21
Burton's Legless Lizard 7, 15, 80, **87, 88**
burtonis 7, 15, 80, **87, 88**

C
calonotus 11, 19, **173, 174**
Cane Toad 21
Caretta caretta 46
Carpet Python 9, 13, **157, 160, 161**
Chelidae 49
Chelodina **50**
  Chelodina oblonga 14, **50**
Chelonia mydas 46, **47**
Cheloniidae **46**
children's pythons **159**
childreni 159
Chlamydosaurus kingii 92
Chocolate Burrowing Frog **37**
christinae **131**
Clawless Gecko **65**
Cobra 196
  Black-necked 162

  Spitting 162
Comb-bearing dragons **94**
Comb-eared Skink **114**
Common Death Adder 12, **164**, 206
Common Dwarf Skink 15, **137**
Common Scaly-foot **90, 91**
concinna **82**
coriacea 46
coronata **168, 169**
Crenadactylus 63, **65**
  Crenadactylus ocellatus 63, **65**
Crinia **29**
  Crinia georgiana **29**
  Crinia glauerti **30**
  Crinia insignifera **31**
  Crinia pseudinsignifera 31, **32**
Crowned Snake **168, 169**
Cryptoblepharus **112**
  Cryptoblepharus plagiocephalus 7, **112**
Ctenophorus **94**
  Ctenophorus ornatus 7, 13, **94, 95**
Ctenotus 106, **114**, 190
  Ctenotus delli 12, **114**, 118
  Ctenotus fallens 9, **115**, 120
  Ctenotus gemmula 11, **116**
  Ctenotus impar **117**
  Ctenotus labillardieri 13, **118, 119**
  Ctenotus lancelini **19**
  Ctenotus lesueurii 11, **115, 120**
Ctenotus
  Darling Range Heath **114**
  Jewelled **116**
  Red-legged **118, 119**
  South-western Odd-striped **117**
  West Coast **115**
  Western Limestone 11, **120**
curta 10, **170, 171**
Cyclodomorphus **121**
  Cyclodomorphus branchialis 10, **121**
Cyclorana 22, 27, 43

INDEX

D
Darling Range Heath Ctenotus 12, 114
Death Adder 12, **164**, 171, 206
delli 12, **114**, 118
***Delma* 55, 85**
  Delma fraseri **85**, 86
  Delma grayii 10, **86**
***Demansia* 57, 166**
  Demansia psammophis **166, 167**
  Demansia psammophis reticulata **166, 167**
Dermochelyidae 46, 47
Dermochelys coriacea 46
diamond pythons **160**
***Diplodactylus* 66, 67**
  Diplodactylus alboguttatus **67**
  Diplodactylus granariensis **68**, 69
  Diplodactylus polyophthalmus 12, 68, **69**
  Diplodactylus pulcher **70**
distinguenda 12, **132**, 133
dorsalis 10, 23, **39**
Dragon **92**
  Ornate Crevice 7, 13, **94, 95**
  Western Bearded **96**
  Western Heath 11, **97**
***Drysdalia* 168**
  Drysdalia coronata **168, 169**
Dtella **71**
Dugite 15, 16, 56, 163, **179-181**, 196, 199, 201, 205, 206

E
earless dragons **97**
earless skinks 9, 12, **128**
Eastern Brown Snake 196
***Echiopsis* 10, 170**
  Echiopsis curta 10, **170, 171**
***Egernia* 9, 106, 122**
  Egernia kingii 9, **122, 123**, 125
  Egernia luctuosa 14, **124**
  Egernia napoleonis 13, **125**
  Egernia pulchra **126**
Egernia
  South-western Crevice **125**
  South-western Spectacled Rock **126**
  Western Glossy Swamp **124**
elegans 15, **133**, 193
***Eremiascincus* 127**

Eremiascincus richardsonii **127**
Eretmochelys imbricata 46
exilis 16, 161, **179, 180**
eyrei 10, 23, 34, **36**, 38

F
fallens 9, **115**, 120
fangs 197
fasciolatus **190**
Fat Blind Snake **154**
Fence Lizard 7, **112**
Fraser's Legless Lizard **85**
fraseri **85**
fresh-water turtles 51
Frilled Lizard 92
Frog
  Chocolate Burrowing **37**
  Glauert's Froglet **30**
  Granite Froglet **32**
  Green-bellied Froglet **33**
  ground 22, 27
  Guenther's Toadlet 14, **42**
  Humming **41**
  Marbled Burrowing **38**
  Moaning 10, 23, **36**
  narrow-mouthed 22
  Pobblebonk 23, **39**
  Red-thighed Froglet **29**
  Sandhill 22
  Sandplain Froglet **31**
  Slender Tree 14, 24, **44**
  Spotted Burrowing **34**
  tree 22, 27
  true 22
  Turtle 11, 22, 23, 24, **40**
  water-holding 22, 27, 43
  Western Banjo 10, 23, **39**
  Western Green Tree 14, 24, **46**
  Yellow-flanked Burrowing **35**
Funnelweb Spider 200

G
Gecko 8, 62
  Barking **78**
  Binoe's Prickly 62, **72**
  Clawless **65**
  Marbled 10, **74**
  Reticulated Velvet 7, **73**
  South-western Spiny-tailed 9, **76, 77**
  Speckled Stone 11, **69**

Thick-tailed **78**
Western Marbled 10
Western Saddled Ground **70**
Wheatbelt Stone **68**, 69
White-spotted Ground 67
***Gehyra* 71**
  Gehyra variegata 63, **71**, 74
Gekkonidae **62**
gemmula 11, **116**
***Geocrinia* 33**
  Geocrinia leai **33**
georgiana **29**
Gila Monster 195
glauerti **30**
Glauert's Froglet **30**
Gould's Monitor 99, 100, **101**
Gould's Snake 13, **186, 187**
Gould's Hooded Snake **186, 187**, 188
gouldii 11, 13, 22, 24, **40**, 99, 100, **101, 186, 187**, 188
gracilis 11, **89**
granariensis **68**, 69
Granite Froglet **32**
Gray's Legless Lizard 10, **86**
grayii 10, **86**
Green Turtle 46, **47**
Green-bellied Froglet **33**
greyii 15, **137**
ground frogs 22, 27, 43
guentheri 14, **42**
Guenther's Toadlet 14, 42
Gwardar **163, 182-185**, 201, 206

H
habitats
  banksia woodland **11**
  coastal limestone & heath **9**
  granite outcrop **13**
  human-made 8, **15**
  jarrah woodland **12**
  offshore islands 8
  tuart woodland **10**
  wetland **14**
Hawksbill Turtle 46, 47
***Heleioporus* 34**
  Heleioporus albopunctatus **34**
  Heleioporus barycragus **35**
  Heleioporus eyrei 10, 23, **34**
  Heleioporus inornatus **37**
  Heleioporus psammophilus **38**
***Hemiergis* 128**

Hemiergis initialis 12, **128**
Hemiergis quadrilineata 9, **129**
Heteronotia 72
Heteronotia binoei 62, **72**
hooded snakes 186
horridus 92, **93**
Humming Frog 41
Hydrophis
Hydrophis elegans 193
Hydrophis kingii 193
Hydrophis major 193
Hydrophis ocellatus 193
Hylidae 22, 27, 43

I
imbricata 46, **160, 161**
impar **117**
initialis 12, **128**
Inland Taipan 196
inornatus **37**, **76**, 77
insignifera **31**

J
Jan's Banded Snake **189**
Javelin Legless Lizard **82**
Jewelled Ctenotus 11, **116**

K
Kabarda **179**
Keeled Legless Lizard 11, **89**
keys, what are? 3
King Brown Snake 178 (see also Mulga Snake)
King's Skink 9, **122, 123**
kingii 9, 92, **122, 123**, 125, 193
Komodo Dragon 195
konowi 16, **141, 142**
krait 196

L
labillardieri 13, **118, 119**
Lacertilia 54, 60
lancelini **19**
Lancelin Island Skink **19**
leai **33**
Leathery Turtle 46
legless lizards **79**
Leiolopisma **110**
lepidopodus **90, 91**
Leptodactylidae 27
Lerista 57, 106, 107, **131**, 172

Lerista christinae **131**
Lerista distinguenda 12, **132**, 133
Lerista elegans 15, **133**
Lerista lineata 18, **134**
Lerista lineopunctulata 9, **135**
Lerista praepedita 10, **136**, 174
Lerista
  Bold-striped **131**
  Perth Lined **134**
  South-western Four-toed **132**
  West Coast Four-toed **133**
  West Coast Line-spotted **135**
  Western Worm **136**, 174
lesueurii 11, 115, **120**
Lialis 79, 80, **87**
Lialis burtonis 7, 15, 80, **87**, **88**
Liasis 159
Liasis olivaceus 156
Liasis olivaceus barroni 156
Limnodynastes 39
Limnodynastes dorsalis 10, 23, **39**
lineata 18, **134**
Line-spotted Lerista 9
lineoocellata 9, **138**, 139
lineopunctulata 9, **135**
Litoria 44
Litoria adelaidensis 14, 24, **44**
Litoria moorei 14, 24, **45**
Lizard (see also dragon, gecko, monitor & skink)
  Burton's Legless 7, 15, **87, 88**
  Fence 7
  Fraser's Legless **85**, 86
  Frilled 92
  Gray's Legless 10, **86**
  Javelin Legless **82**
  Keeled Legless 11, **89**
  Mexican Beaded 195
  Sleepy **141**
  South-western Sandplain Worm 11, **84**
  Western Granite Worm **83**
Loggerhead 46
Long-necked Turtle **50**
luctuosa 14, **124**
Luth 46

M
major 193
Marbled Burrowing Frog **38**
Marbled Gecko 10, **74**

marinus 21
Marine Toad 21
marmoratus 10, 63, 71, 73, **74**, **75**
Menetia 107, **137**
Menetia greyii 15, **137**
Mexican Beaded Lizard 195
Microhylidae 22
milii **78**
minor **96**
Moaning Frog 10, 23, **36**, 38
Moloch horridus 92, **93**
Monitor
  Black-tailed **104, 105**
  Gould's 99, **101**
  Short-tailed 99
  Southern Heath 100, **102, 103**
moorei 14, 24, **45**
Morelia (also see *Antaresia*) 159, **160**
  Morelia amethistina 156
  Morelia spilota 9, 13, **160, 161**
  Morelia spilota imbricata **160, 161**
Morethia 107, **138**
  Morethia lineoocellata 9, **138**, 139
  Morethia obscura 12, 138, **139**
Morethia
  Southern Pale-flecked **139**
  Western Pale-flecked **138**
Mulga Snake **177, 178**, 206
mydas 46, **47**
Myobatrachidae 22, 27, 43
Myobatrachus 40
  Myobatrachus gouldii 11, 22, 24, **40**

N
names, scientific 4
napoleonis 13, **125**
Narrow-banded Snake **190**
narrow-mouthed frogs 22
Neelaps 172
  Neelaps bimaculatus **172**
  Neelaps calonotus 11, 19, **173, 174**
Neobatrachus 41
  Neobatrachus pelobatoides **41**
nigriceps 13
Northern Taipan 196, 200
Notechis 175, 200
  Notechis scutatus 14, **175, 176**, 206

Notechis scutatus occidentalis
    175, 176
nuchalis **182-185**, 201, 206

O
oblonga 14, **50**
Oblong Turtle 14, **50**
obscura 12, 138, **139**
occidentalis **175, 176**
occipitalis **140**, 142
ocellatus 63, **65**, 193
Oedura 7, **73**
    Oedura reticulata 7, **73**
olivaceus 156
Olive-headed Sea Snake 193
Ornate Dragon 7, 13, **94**
Ornate Crevice Dragon 13, **94, 95**
ornatus 7, 13, **94, 95**
Oxyuranus scutellatus 200

P
Pelamis platurus 192, 193, **194**
Perentie 99, 195
Perth Lined Lerista 18, **134**
perthensis 156
Phyllodactylus 74
    Phyllodactylus marmoratus 10,
        63, 71, 73, **74**, **75**
Pilbara Olive Python 156
pinguis **154**
plagiocephalus 7, **112**
platurus 192, 193, **194**
Pletholax 55, **89**
    Pletholax gracilis 11, **89**
Pobblebonk 23, **39**
Pogona **96**
    Pogona minor **96**
polyophthalmus 12, 68, **69**
pooleorum 193
praepedita 10, **136**, 174
psammophis **166, 167**
psammophilus 38
Pseudechis **177**
    Pseudechis australis **177, 178**,
        206
    Pseudechis papuanus 177
Pseudemydura **52**
    Pseudemydura umbrina 18, **52**
Pseudonaja **179**, 200
    Pseudonaja affinis 15, **179-181**,
        206

Pseudonaja affinis exilis 16, 161,
    179, 180
Pseudonaja nuchalis **182-185**,
    201, 206
Pseudophryne **42**
    Pseudophryne guentheri 14, **42**
pulchella 13, **83**, 84
pulcher **70**
pulchra **126**
Pygmy Python 156
Pygopodidae 55, **79**
Pygopus 55, **90**
    Pygopus lepidopodus **90, 91**
Python
    Amethyst 156
    Carpet 9, 13, **160, 161**
    Children's 159
    Diamond **160**
    Pilbara Olive 156
    Pygmy 156
    Sand **20**
    Scrub 156
    Stimson's **159**

Q
quadrilineata 9, **129**

R
Ramphotyphlops **149**
    Ramphotyphlops australis 12,
        150, **152, 153**
    Ramphotyphlops pinguis 150, **154**
    Ramphotyphlops waitii 150, **155**
ramsayi **20**
Ranidae 22
Rankinia 97
Red-legged Ctenotus 13, **118**
Red-thighed Frog **29**
repens 11, 83, **84**, 174
reticulata 7, **73**, **166**, 167
Reticulated Velvet Gecko 7, **73**
Reticulated Whip Snake **166, 167**
Rhinoplocephalus **186**
    Rhinoplocephalus gouldii 13, **186,
        187**
    Rhinoplocephalus nigriceps 13,
        186, **188**
richardsonii **127**
Ringhals 162
rosenbergi 100, **102, 103**
rotunda 22

rugosa 15, 106, **141, 142**
Rusty Gecko 12

S
Sand Python **20**
Sandhill Frog 22
Sandplain froglet **31**
Sandplain Worm Lizard 174
sandswimming skinks **131**
scientific names, what are? 4
Scrub Python 156
scutatus 14, **175, 176**, 206
Sea Snake
    Bar-bellied 193
    Olive-headed 193
    Shark Bay 193
    Spectacled 193
    Spotted 193
    Yellow-bellied 192, 193, **194**
sea turtles 46
semifasciatus **191**
Shark Bay Sea Snake 193
Shingleback **141, 142**
Short-necked Turtle **52**
Short-tailed Monitor 99
Simoselaps 9, 172, **189**
    Simoselaps bertholdi **189**
    Simoselaps fasciolatus **190**
    Simoselaps semifasciatus **191**
Skink 8, **106** (see also egernia,
    lerista and morethia)
    Bluetongue 10, **140**, 142
    Bobtail 3, 15, 16, 106, **141, 142**
    Comb-eared **114**
    Common Dwarf 15, **137**
    Earless 9, 12
    Fence 7, **112**
    King's 9, **122, 123**
    Lancelin Island **19**
    Sandswimming **127**
    Snake-eyed **112**
    Southern Five-toed Earless 12,
        128
    South-western Cool **110, 111**
    Sun **112**
    Two-toed Earless 9
Sleepy Lizard (see also Bobtail)
Slender Tree Frog 14, 24, **44**
Snake (see also Sea Snake)
    Banded Sand **189**
    Beaked Blind **155**

Black-backed 13, 186, **188**
Black-naped **172**
Black-striped 11, 19, **173, 174**
Crowned **168, 169**
Eastern Brown 196
Fat Blind **154**
Gould's Hooded 13, **186, 187**
Half-girdled **191**
Jan's Banded **189**
King Brown **178**
Mulga **177, 178**
Narrow-banded **190**
Reticulated Whip **166**
Southern Blind 12, **152, 153**
Southern Half-girdled **191**
Spotted Brown **179-181**, 199, 205
Tiger 14, 56, **175**, 196
Western Brown **182-185**, 196, 201
Western Tiger **175, 176**, 199 206
South-western Cool Skink 14, **110, 111**
South-western Crevice Egernia 13, **125**
South-western Four-toed Lerista 12, **132**
South-western Odd-striped Ctenotus **117**
South-western Sandplain Worm Lizard 11, **84**
South-western Spectacled Rock Egernia **126**
South-western Spiny-tailed Gecko **76, 77**
Southern Blind Snake 12, **152, 153**
Southern Carpet Python 13, **160, 161**
Southern Five-toed Earless Skink 12, **128**
Southern Half-girdled Snake **191**
Southern Heath Monitor 100, **102, 103**
Southern Pale-flecked Morethia 12, 138, **139**
Speckled Stone Gecko **69**
Spectacled Sea Snake 193
spilota 9, **160, 161**
spinigerus 9, **76, 77**
Spotted Brown Snake **179-181**, 199, 205
Spotted Burrowing Frog **34**
Spotted Sea Snake 193

Squamata **54**, 60
stimsoni **159**
Stimson's Python **159**
Strophurus 9, **76**
   Strophurus spinigerus 9, **76, 77**
Sun Skink **112**

T

Taipan
   Inland 196
   Northern 196, 200
Thick-tailed Gecko **78**
Thorny Devil 92, **93**
tiger snakes 14, 56, **175**
***Tiliqua* 140**
   Tiliqua occipitalis **140**, 142
   Tiliqua rugosa 15, 106, **141, 142**
Toad
   Cane 21
   Marine 21
Toadlet
   Guenther's 14, **42**
Tortoise (see Turtle)
tree frogs 22
trilineata **110, 111**
tristis **104, 105**
true frogs 22
Turtle
   freshwater 49
   Green 46, **47**
   Hawksbill 46, 47
   Leathery 46
   Loggerhead 46
   Long-necked **50**
   Luth 46
   Oblong 14, **50**
   sea 46
   Short-necked **52**
   Western Swamp **52**
Turtle Frog 22, 23, 24, **40**
Two-toed earless Skink 9, **129**
***Tympanocryptis* 97**
   Tympanocryptis adelaidensis 11, **97, 98**

U

umbrina 18, **52**
Underwoodisaurus 78
   Underwoodisaurus milii **78**

V

***Varanus* 99**
   Varanus brevicauda 99
   Varanus giganteus 99
   Varanus gouldii 99, 100, **101**
   Varanus rosenbergi 100, **102, 103**
   Varanus tristis **104, 105**
Variegated Dtella **71**
variegata 63, **71**, 74
Vermicella **172**
viper 196

W

waitii **155**
water-holding frogs 22, 27, 43
West Coast Ctenotus 9, **115**
West Coast Four-toed Lerista 15, **133**
West Coast Line-spotted Lerista **135**
Western Banjo Frog 10, 23, **39**
Western Bearded Dragon **96**
Western Bluetongue **140**, 142
Western Brown Snake **182-185**, 201, 206
Western Glossy Swamp Egernia 14, **124**
Western Granite Worm Lizard 13, **83**
Western Green Tree Frog 14, 24, **45**
Western Heath Dragon 11, **97, 98**
Western Limestone Ctenotus 11, **120**
Western Marbled Gecko 10
Western Pale-flecked Morethia **138**, 139
Western Saddled Ground Gecko **70**
Western Slender Bluetongue 10, **121**
Western Swamp Turtle 18, **52**
Western Tiger Snake **175, 176**, 199, 206
Western Worm Lerista 10, **136**, 174
Wheatbelt Stone Gecko **68**
whip snakes 57, **166**
white-lipped snakes **168**
White-spotted Ground Gecko **67**
Woma **20**
worm lizards 11, 12, **83**
worm snakes **149**

Y

Yellow-bellied Sea Snake 192, 193, **194**
Yellow-flanked Burrowing Frog **35**

First aid for snakebite

(See next page for diagram.)

Assume all snakes are venomous. Always believe someone when they say they have been bitten by a snake, even though you may not see any puncture marks.

Symptoms: nausea, sweating, diarrhoea, pains in the chest, double vision.

Signs: puncture marks, slight bruising, redness, swelling.

Rest the casualty. Do not panic. Apply direct pressure over the bitten area. Apply a firm broad bandage over the bite area first, then bandage down the limb and continue to bandage up the full length of the limb. Immobilise the limb with a splint. Call for medical aid.

If bandages and splint have been applied correctly, they will be comfortable and may be left on for several hours.

They should not be taken off until the patient has reached medical care. The doctor will decide when to remove the bandages.

If venom has been injected, it will move into the system very quickly when the bandages are removed. The doctor should leave them in position until he or she has assembled appropriate antivenom and drugs which may have to be used when the dressings and splint are removed.

First aid for snakebite

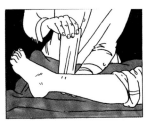
1.

2.

3.

4.

5.

6.

1. Apply a broad pressure bandage over the bite site as soon as possible (don't take off jeans as the movement of doing so will assist the venom to enter the bloodstream. Keep the bitten leg still!).
2. The bandage should be as tight as you would apply to a sprained ankle.
3. Extend the bandages as high as possible.
4. Apply a splint to the leg.
5. Bind it firmly to as much of the leg as possible.
6. Bites on hand or forearm: Bind to elbow with bandages; Use splint to elbow; Use sling.

Based on material by Dr S K Sutherland, Commonwealth Serum Laboratories, Parkville, Victoria (1985).